U.S. Department of Transportation
National Highway Traffic Safety Administration

DOT HS 811 538

November 2011

Evaluation of the 1999-2003 Head Impact Upgrade of FMVSS No. 201 Upper-Interior Components

Effectiveness of Energy-Absorbing Materials Without Head-Protection Air Bags

DISCLAIMER

This publication is distributed by the U.S. Department of Transportation, National Highway Traffic Safety Administration, in the interest of information exchange. The opinions, findings, and conclusions expressed in this publication are those of the authors and not necessarily those of the Department of Transportation or the National Highway Traffic Safety Administration. The United States Government assumes no liability for its contents or use thereof. If trade names, manufacturers' names, or specific products are mentioned, it is because they are considered essential to the object of the publication and should not be construed as an endorsement. The United States Government does not endorse products or manufacturers.

Suggested APA Format Reference:

Kahane, C. J. (2011, November). Evaluation of the 1999-2003 head impact upgrade of FMVSS No. 201 – Upper-interior components: Effectiveness of energy-absorbing materials without head-protection air bags. (Report No. DOT HS 811 538). Washington, DC: National Highway Traffic Safety Administration.

Technical Report Documentation Page

1. Report No.	2. Government Accession No.	3. Recipient's Catalog No.
DOT HS 811 538		
4. Title and Subtitle		5. Report Date
Evaluation of the 1999-2003 Head Impact Upgrade of FMVSS No. 201 – Upper-Interior Components: Effectiveness of Energy-Absorbing Materials Without Head-Protection Air Bags		November 2011
		6. Performing Organization Code
7. Author(s)		8. Performing Organization Report No.
Charles J. Kahane, Ph.D.		
9. Performing Organization Name and Address		10. Work Unit No. (TRAIS)
Office of Vehicle Safety National Highway Traffic Safety Administration Washington, DC 20590		
		11. Contract or Grant No.
12. Sponsoring Agency Name and Address		13. Type of Report and Period Covered
National Highway Traffic Safety Administration 1200 New Jersey Avenue SE. Washington, DC 20590		NHTSA Technical Report
		14. Sponsoring Agency Code
15. Supplementary Notes		
16. Abstract		
Federal Motor Vehicle Safety Standard (FMVSS) No. 201 – Occupant Protection in Interior Impact – was upgraded in 1995, with a 1998-2003 phase-in, to reduce occupants' risk of head injury from contact with a vehicle's upper interior, including its pillars, roof headers and side rails, and the upper roof. Initially, energy-absorbing materials alone were used to meet the standard. NHTSA statistically analyzed the effect of these materials on head injuries due to upper-interior contact in cars and light trucks in the Crashworthiness Data System of the National Automotive Sampling System for 1995-2009 and the effect on head injuries in fatal crashes in the Fatality Analysis Reporting System – Multiple Cause of Death files for 1999-2007. FMVSS No. 201 without head-protection air bags reduces AIS 4-to-6 head injuries due to contact with upper-interior components by an estimated 24 percent (95% confidence bounds, 11 to 35%), based on the average of the analysis results for the two databases. That is equivalent to a 4.3-percent reduction of overall fatality risk (confidence bounds 2.0 to 6.2%). When all vehicles on the road meet FMVSS No. 201, it will save an estimated 1,087 to 1,329 lives per year. At a cost of $25.52 (in 2010 dollars) over the life of a vehicle, that amounts to an annual cost, depending on new-vehicle sales, ranging from $301 to $424 million for certifying all new vehicles to FMVSS No. 201. It is a very cost-effective regulation, costing less than $1 million per life saved.		
17. Key Words	18. Distribution Statement	
NHTSA; NASS; CDS; FARS; MCOD; FMVSS; HIC; crashworthiness; concussion; A-pillar; B-pillar; head impact; Rao-Scott chi-square	Document is available to the public from the National Technical Information Service www.ntis.gov	
19. Security Classif. (Of this report)	20. Security Classif. (Of this page)	21. No. of Pages	22. Price
Unclassified	Unclassified	122	

Form DOT F 1700.7 (8-72) Reproduction of completed page authorized

TABLE OF CONTENTS

List of abbreviations iii

Executive summary v

1. Head protection in impacts with a vehicle's upper interior 1
   1.1 The problem: head injuries due to upper-interior contact 1
   1.2 The head-injury protection upgrade for FMVSS No. 201 9
   1.3 In what ways were vehicles modified? 10
   1.4 HIC test results before and after FMVSS No. 201 17
   1.5 Fatality reduction by head-protection air bags in side impacts 19

2. Effect of FMVSS No. 201 on AIS 3-to-6 head injuries from upper-interior sources: analyses of 1995-2009 CDS data 23
   2.0 Summary 23
   2.1 A database of AIS 3-to-6 injuries before and after FMVSS No. 201 23
   2.2 Overall reduction of AIS 3-to-6 head injuries from upper-interior sources 26
   2.3 Effect on AIS 4 to 6 for specific impact, vehicle, occupant, and injury types 29
   2.4 Effect on AIS 3 to 6 for specific impact, vehicle, occupant, and injury types 33

3. Effect of FMVSS No. 201 on head injuries in fatal crashes: analyses of 1999-2007 FARS-MCOD data 37
   3.0 Summary 37
   3.1 Injuries contributing to occupant fatalities, before and after FMVSS No. 201 37
   3.2 Overall reduction of head injuries 41
   3.3 Belted versus unrestrained occupants 43
   3.4 Effect for specific impact, vehicle, and occupant types 45

4. Discussion: best effectiveness estimate, benefits, and costs 48
   4.0 Summary 48
   4.1 Best effectiveness estimate and its confidence bounds 48
   4.2 Effect on fatalities 49
   4.3 Lives saved and potentially savable by FMVSS No. 201 in CY 2004-2009 50
   4.4 Cost of FMVSS No. 201; cost per life saved 54
   4.5 Upper-interior head injuries after FMVSS No. 201/head-protection air bags 57

Appendix A: Initial model year of FMVSS No. 201 certification: makes and models produced in 1999-2003 60
Appendix B: Listing and classification of ICD-10 S and T codes 65
Appendix C: Make-model groups for evaluating FMVSS No. 201: model-year ranges included in the FARS-MCOD analyses 74

LIST OF ABBREVIATIONS

AIS	Abbreviated Injury Scale
BEA	Bureau of Economic Analysis, U.S. Department of Commerce
BMW	Bayerische Motoren Werke
CAC	Certified advanced compliant air bag
CDC	Centers for Disease Control and Prevention, U.S. Department of Health and Human Services
CDS	Crashworthiness Data System of NASS
CG	Car group, vehicle group (from NHTSA's VIN analysis programs)
CY	Calendar year
DEFF	Design effect
DOT	United States Department of Transportation
EIA	Energy Information Administration, U.S. Department of Energy
ESC	Electronic stability control
FARS	Fatality Analysis Reporting System, a census of fatal crashes in the United States since 1975
FMH	Free-motion headform
FMVSS	Federal Motor Vehicle Safety Standard
FRIA	Final Regulatory Impact Analysis
GAD	General area of damage
GDP	Gross domestic product
GMC	General Motors Corporation
GVWR	Gross vehicle weight rating, specified by the manufacturer, equals the vehicle's curb weight plus maximum recommended loading
HIC	Head Injury Criterion
ICD	International Classification of Diseases

ICD-10	International Classification of Diseases, 10th revision
LTV	Light trucks and vans, includes pickup trucks, SUVs, minivans, and full-size vans
LWB	Long wheelbase
MCOD	Multiple cause of death file, a part of FARS
MM2	Make-model code (from NHTSA's VIN analysis programs)
mph	Miles per hour
MY	Model year
NASS	National Automotive Sampling System, a probability sample of police-reported crashes in the United States since 1979, investigated in detail
NCHS	National Center for Health Statistics
NHTSA	National Highway Traffic Safety Administration, U.S. Department of Transportation
NPRM	Notice of Proposed Rulemaking
NPV	Net present value
PSU	Primary sampling unit
RATWGT	[Inverse sampling] ratio weight
RF	Right-front seat
SAE	Society of Automotive Engineers
SAS	Statistical and database management software produced by SAS Institute, Inc.
SRS	Simple random sampling
SUV	Sport utility vehicle
VIN	Vehicle Identification Number
VMT	Vehicle miles of travel
VSL	Value of a statistical life
WHO	World Health Organization

EXECUTIVE SUMMARY

The purpose of Federal Motor Vehicle Safety Standard (FMVSS) No. 201 – Occupant Protection in Interior Impact – is to reduce occupants' risk and the severity of head injury in crashes. The performance-test requirements of FMVSS No. 201 limit the force allowed when a dummy headform impacts locations in the vehicle's interior that might be contacted by occupants' heads during crashes. NHTSA's major upgrade of FMVSS No. 201 in 1995 added the A-, B- and other pillars; roof headers; roof side rails; and the upper roof to the list of test locations. The intensity and duration of the force at the new test locations is measured by the head injury criterion, which may not exceed 1,000 for any 36-millisecond period during a 15-mph headform impact. Extensive laboratory testing and biomechanics research by NHTSA and other organizations had demonstrated that high HIC is associated with a high risk of head injury and low HIC with low risk.

The new requirements phased into passenger cars and LTVs (pickup trucks, SUVs and vans) with less than 10,000 pounds gross vehicle weight rating during 1998-2003. Initially, energy-absorbing materials alone were used to meet the standard. The materials – located beneath the potential impact locations – include composite plastic foam padding, injection-molded ribs or ridges in parallel or egg-crate-like configurations, crushable tubes, and more flexible designs for interior surfaces and components. NHTSA issued a study in 2006 showing that these materials had reduced average HIC in the 15-mph headform impacts from 910 in pre-standard vehicles to 668 in the post-standard vehicles – a statistically significant average improvement of 242 units of HIC.[1]

More recently, increasing numbers of new cars and LTVs have also been equipped with curtain air bags or other types of head-protection air bags – e.g., 66 percent of model-year 2007 vehicles. NHTSA anticipates that all new vehicles will be equipped with head-protection air bags by September 1, 2014, when an upgrade to FMVSS No. 214 (Side Impact Protection) adding a side-impact with a pole goes into full effect.

This report is a statistical evaluation of the fatality- and injury-reducing effectiveness of the energy-absorbing materials in vehicles without head-protection air bags. (NHTSA previously evaluated the effectiveness of head-protection air bags in 2007.[2]) In one sense, this report evaluates a specific technological approach (energy-absorbing materials without air bags) that is already phasing out. But the energy-absorbing materials, themselves, will not be phasing out; they will continue to appear in new vehicles to protect occupants in crashes where the air bags do not deploy or perhaps at locations not covered by the air bags. More generally, the report investigates whether a technology demonstrated to have reduced HIC measured on headforms in laboratory testing is likewise effective in reducing the head injuries of people in crashes.

---

[1] Kahane, C. J., & Tarbet, M. J. (2006, November). *HIC test results before and after the 1999-2003 head impact upgrade of FMVSS 201.* (Report No. DOT HS 810 739). Washington, DC: National Highway Traffic Safety Administration. Available at www-nrd.nhtsa.dot.gov/Pubs/810739.PDF.

[2] Kahane, C. J. (2007, January). *An evaluation of side impact protection – FMVSS 214 TTI(d) improvements and side air bags.* (Report No. DOT HS 810 748). Washington, DC: National Highway Traffic Safety Administration. Available at www-nrd.nhtsa.dot.gov/Pubs/810748.PDF.

The statistical analyses are based on the Crashworthiness Data System of the National Automotive Sampling System for 1995-2009 and the Fatality Analysis Reporting System – Multiple Cause of Death (FARS-MCOD) files for 1999-2007. CDS lists each individual injury for crash-involved occupants and specifies its source (contact point within the vehicle), the type of injury, and its severity, based on the Abbreviated Injury Scale. With CDS, it is possible to find the reduction, after FMVSS No. 201, of life-threatening (AIS 4 to 6) or serious (AIS 3 to 6) head injuries due to contact with upper-interior target areas such as the A-pillar or roof interior – relative to a control group of the injuries that are not head injuries and/or do not involve upper-interior contact. FARS-MCOD lists, for fatally injured people, the body regions (and sometimes other details) of injuries that, according to the person's death certificate "contributed to the fatality." It does not specify injury sources within the vehicle. With FARS-MCOD, it is possible to find the reduction, after FMVSS No. 201, of head injuries relative to a control group of injuries to other body regions.

The report's principal findings and conclusions are the following:

INJURY RISK BEFORE THE FMVSS No. 201 UPGRADE

- Just before FMVSS No. 201, head injuries due to upper-interior contact accounted for 22 percent of all AIS 4-to-6 injuries and 14 percent of all AIS 3-to-6 injuries in the United States.

- Upper-interior contacts accounted for 43 percent of all AIS 4-to-6 head injuries and 42 percent of AIS 3-to-6 head injuries.

- The upper roof, B-pillar, A-pillar, and roof side rail account for the largest numbers of head injuries due to upper-interior contact.

- The vast majority of the AIS 4+ as well as AIS 3+ head injuries due to upper-interior contact are brain injuries.

- An estimated 18 percent of occupant fatalities could have been prevented by preventing all head injuries due to upper-interior contact.

CDS ANALYSIS

- After FMVSS No. 201 certification (without head-protection air bags), life-threatening (AIS 4 to 6) head injuries due to upper-interior contact decrease by a statistically significant 32 percent. For belted occupants, the reduction is a statistically significant 36 percent.

- The reduction of serious (AIS 3 to 6) head injuries due to upper-interior contact is a non-significant 11 percent.

- FMVSS No. 201 is especially effective in reducing life-threatening injuries in first-event rollovers, in head impacts with the A-pillar or roof interior (where HIC was most strongly reduced in headform tests), and against concussions.

- Observed effectiveness did not differ substantially between cars and LTVs, by occupant seat position, by occupant age, or between males and females.

FARS-MCOD ANALYSIS

- Immediately after FMVSS No. 201 certification (without head-protection air bags), head injuries of fatally injured occupants decrease by a statistically significant 6 percent relative to injuries to other body regions. For belted occupants, the reduction is a statistically significant 10 percent.

- Given that upper-interior contacts accounted for 43 percent of life-threatening head injuries (see above), the preceding 6- and 10-percent overall reductions of head injuries may imply 15- and 23-percent reductions of head injuries due to upper-interior contact.

- FMVSS No. 201 is especially effective in reducing head injuries in first-event rollovers (consistent with CDS), in LTVs, and for female occupants (two trends not seen in CDS).

- Observed effectiveness did not differ substantially by occupant seat position or by occupant age (consistent with CDS).

BEST EFFECTIVENESS ESTIMATE, BENEFITS, AND COSTS

- NHTSA estimates FMVSS No. 201 without head-protection air bags reduces AIS 4-to-6 head injuries due to contact with upper-interior components by 24 percent (95% confidence bounds, 11 to 35%). It is the average of the CDS estimate (32%) and the reduction of head injuries due to upper-interior contact implied by the FARS analysis (15%).

- Given that 18 percent of occupant fatalities could have been prevented by preventing all head injuries due to upper-interior contact (see above), the preceding 24-percent reduction is equivalent to a 4.3-percent reduction of overall fatality risk (confidence bounds 2.0 to 6.2%).

- If occupant fatalities in calendar years 2004-2007 are considered a baseline, an on-road fleet entirely certified to FMVSS No. 201 will save an estimated 1,329 lives per year (confidence bounds 621 to 1,925) relative to an entirely pre-standard on-road fleet. With the lower baseline of 2008-2009 fatalities, FMVSS No. 201 will save an estimated 1,087 lives per year (confidence bounds 500 to 1,575).

- FMVSS No. 201 without head-protection air bags adds, on the average, $25.52 (in 2010 dollars) to the lifetime cost of purchasing and operating a car or LTV. The operating cost comprises added fuel consumption due to an average 2.84 pounds of added weight.

- At 2004-2007 sales levels for new cars, that amounts to an annual cost of $424 million, in 2010 dollars. At 2008-2009 sales levels, the annual cost would be $301 million.

- Even if vehicle sales should return to 2004-2007 levels while baseline fatalities continue, hopefully, to stay near the 2008-2009 levels, the estimated cost per life saved is only $390,000 (in 2010 dollars, confidence bounds $269,000 to $848,000). The energy-absorbing materials are very cost-effective, costing far less than the Department of Transportation's estimate of $6,000,000 to $6,300,000 for the value of a statistical life.

CHAPTER 1

HEAD PROTECTION IN IMPACTS WITH A VEHICLE'S UPPER INTERIOR

Federal Motor Vehicle Safety Standard (FMVSS) No. 201 – Occupant Protection in Interior Impact – was upgraded in 1995, with a model year 1999-2003 phase-in, to reduce occupants' risk and severity of head injury from contact during crashes with a vehicle's upper interior, including its pillars, roof headers and side rails, and the upper roof. Initially, energy-absorbing materials alone were used to meet the standard; later, more and more vehicles were also equipped with head-protection air bags such as curtain air bags. NHTSA has published a report showing that the energy-absorbing materials have effectively reduced the Head Injury Criterion in laboratory tests[3] and another report showing that head-protection air bags save lives in actual crashes.[4] Crash data is now available to evaluate the injury-reducing effectiveness of the energy-absorbing materials alone.

1.1 The problem: head injuries due to upper-interior contact

The Crashworthiness Data System of the National Automotive Sampling System is a national probability sample since 1979 of passenger cars and LTVs (pickup trucks, SUVs and vans with less than 10,000 pounds GVWR) involved in crashes where at least one vehicle was towed from the scene. CDS data from 1995 to 2009 use the same definitions for coding injuries and their sources (contact points). They are an excellent database for analyzing the safety problem – head injuries due to upper interior contact – before the upgrade of FMVSS No. 201.

CDS documents occupants' injuries based on data from hospitals, treatment facilities and autopsies, to the extent that information is available. The analyses consider injuries rated 3 to 6 on the Abbreviated Injury Scale:

3   Serious (but not life-threatening),

4   Severe (life-threatening, survival probable),

5   Critical (life-threatening, survival uncertain), and

6   Not survivable with current medical technology.

An occupant may have more than one such injury. CDS documents the body region of the injury, the lesion and the system or organ involved. In the tables that follow, AIS 3-to-6 injuries are classified into four groups. All burns (as defined by the lesion, regardless of the body region) constitute one group. Except for burns, the injuries with body region H (head) or F (face) constitute the "head injury" group. Neck injuries (body region N) are a third group; they may be triggered by a contact with the upper interior, but less likely a direct contact. Injuries to all other known body regions, including the chest, abdomen, back, arms, and legs are the fourth group, far less likely to involve contact with the upper interior. Injuries with body regions O (whole body)

---

[3] Kahane & Tarbet (2006).
[4] Kahane (2007).

or U (unknown) are excluded from the tables, because it is unclear if they include head injuries or not.

The injury source (contact point) is the other factor in the tables. The following injury source codes comprise the "upper interior" of the vehicle:

>A-pillar (Injury Source Codes 53 and 103),
>B-pillar (54 and 104),
>Other pillar (55 and 105),
>Front header (201),
>Front header plus sun visor (20),
>Rear header (202),
>Roof side rail (203 and 204),
>Roof interior, roof maplite, sunroof (205, 206 and 207),
>Windshield plus surrounding structures (A-pillar and/or front header) (15 and 16), and
>Side window plus surrounding structures (pillars and/or roof side rail) (59 and 109).

The last two are included because they involve upper interior structures (pillars, headers, roof side rails) even though they also involve other components (glazing). All other known injury source codes, including 601 (fire in vehicle) and 603 (non-contact injury), are not considered to be part of the upper interior. Injuries of unknown source (codes 697 or blank) are excluded from the tables.

Table 1-1W shows the distribution of AIS 3-to-6 injuries in the United States, based on weighted 1995-2009 CDS data.[5] Table 1-1W explores the safety problem as it existed **just before** the upgrade of FMVSS No. 201. For that reason, it does not include every car and LTV on CDS, but **excludes** injuries in:

- Vehicles certified to the upgraded FMVSS No. 201;

- Occupants at a seat position equipped with head-protection air bags (even if the vehicle is not certified to the upgraded FMVSS No. 201);

- Drivers or right-front passengers in frontal crashes if the position was not equipped with frontal air bags (because frontal air bags had become standard before the FMVSS No. 201 upgrade phased in);

- Occupants in side impacts if the position was equipped with side air bags for torso protection (because torso bags were only available in relatively few vehicles before FMVSS No. 201 phased in);

- Drivers or right-front passengers at seats equipped with 2-point automatic seat belts (a technology discontinued several years before the FMVSS No. 201 phase-in); and

- Occupants not located at a designated seating position – e.g., riding in the bed of a pickup truck – because they would not eventually be covered by FMVSS No. 201.

---

[5] After Table 1-1W, corresponding results will also be tabulated for unweighted CDS data. When the same table is shown for weighted and unweighted data, we will designate the weighted data table with a W and the unweighted data table with a U.

There are 25,661 individual cases of AIS 3-to-6 injuries with known contact source, meeting the above criteria, in the 1995-2009 CDS files. CDS, however, is not merely a collection of individual cases, but a probability sample of the Nation's towaway crashes. Each case has a ratio weight factor equal to the inverse of its probability of selection. For unbiased national estimates of totals, each case needs to be weighted by RATWGT. The data cells in Table 1-1W are the sum of the RATWGTs for the cases in that cell, rather than just a case count.

TABLE 1-1W

INJURIES BY BODY REGION AND CONTACT SOURCE BEFORE FMVSS No. 201
(Weighted 1995-2009 CDS, rounded to the nearest integer)

	Upper Interior	Not Upper Interior	Total
AIS 3 to 6			
Head injuries	229,346	315,787	545,133
Neck injuries	43,371	41,487	84,569
Torso, arms or legs	62,859	934,978	997,837
Burns	0	5,728	5,728
	335,483	1,297,981	1,633,557
AIS 4 to 6			
Head injuries	103,865	135,669	239,534
Neck injuries	5,867	7,189	13,056
Torso, arms or legs	13,295	201,222	214,517
Burns	0	5,300	5,300
	123,028	349,379	472,407

Just before FMVSS No. 201, head injuries due to upper interior contact accounted for 14 percent (229,346/1,633,557) of all AIS 3-to-6 injuries in the United States. Head injuries due to upper interior contact accounted for 42 percent (229,346/545,133) of all AIS 3-to-6 head injuries. By contrast, only 6 percent of AIS 3-to-6 injuries to torso, arms or legs were attributed to the upper interior (62,859/997,837).

Among life-threatening (AIS 4 to 6) injuries, head injuries due to upper interior contact accounted for a substantially higher 22 percent (103,865/472,407) of all AIS 4-to-6 injuries in the United States, and 43 percent (103,865/239,534) of AIS 4-to-6 head injuries. Among the

229,346 AIS 3-to-6 head injuries due to upper interior contact, 45 percent (103,865) are life-threatening.

Although weighted CDS data is needed for unbiased estimates, rates based on unweighted data (simple case counts) can be more precise in more detailed analyses of small subsets of the data. Table 1-1U is the overall distribution of the 25,661 actual CDS injury cases.

TABLE 1-1U

INJURIES BY BODY REGION AND CONTACT SOURCE BEFORE FMVSS No. 201
(Unweighted 1995-2009 CDS)

	Upper Interior	Not Upper Interior	Total
AIS 3 to 6			
Head injuries	4,232	6,024	10,256
Neck injuries	467	559	1,026
Torso, arms or legs	729	13,551	14,280
Burns	0	99	99
	5,428	20,233	25,661
AIS 4 to 6			
Head injuries	1,995	2,884	4,879
Neck injuries	97	146	243
Torso, arms or legs	233	3,762	3,995
Burns	0	85	85
	2,325	6,877	9,202

Here, head injuries due to upper interior contact accounted for 41 percent (4,232/10,256) of all AIS 3-to-6 head injuries. That is almost identical to the 42 percent in the weighted data.

Table 1-2 reveals that the upper roof, B-pillar, A-pillar and roof side rail account for the largest numbers of head injuries due to upper-interior contact, at both the AIS 3-to-6 and AIS 4-to-6 levels. Injuries may involve surfaces near the front of the upper interior (front header), the front-side (A-pillar), the sides (B-pillar, other pillar), the rear (rear header), the side-top (roof side rail), and the top (roof interior). This suggests that FMVSS No. 201 has the potential to mitigate injuries in planar impacts from every direction (front, left, right, rear) and in rollovers, too. Table 1-3 shows the vast majority of the AIS 3+ as well as AIS 4+ head injuries due to upper interior contact are brain injuries. CDS often has no details on the type of brain injury. Skull fractures are far less common; facial fractures are relatively rare at AIS 3 and nonexistent at AIS 4 to 6.

TABLE 1-2

UPPER-INTERIOR SOURCES OF HEAD INJURIES BEFORE FMVSS No. 201
(Unweighted 1995-2009 CDS)

	AIS 3 to 6		AIS 4 to 6	
	N	%	N	%
Front header, sun visor	425	10.4	190	9.5
Windshield + A-pillar and/or front header	215	5.1	111	5.6
A-pillar	719	17.0	307	15.4
B-pillar	971	22.9	445	22.3
Other pillar	120	2.8	48	2.4
Roof side rail	614	14.5	300	15.0
Side window + pillars and/or roof side rail	24	.6	13	.7
Rear header	44	1.0	18	.9
Roof interior, maplite, sunroof	1,100	26.0	563	28.2
	4,232	100.0	1,995	100.0

TABLE 1-3

UPPER-INTERIOR HEAD INJURIES BEFORE FMVSS No. 201, BY INJURY TYPE
(Unweighted 1995-2009 CDS)

	AIS 3 to 6		AIS 4 to 6	
	N	%	N	%
Concussion	346	8.2	202	10.1
Brain contusion	536	12.7	116	5.8
Unknown brain injury	2,388	56.4	1,210	60.7
Skull fracture	614	14.5	227	11.4
Facial fractures	100	2.4	none	.
All other head injuries	248	5.9	240	12.0
	4,232	100.0	1,995	100.0

Key statistics are the percentages of AIS 3-to-6 and AIS 4-to-6 head injuries that are due to upper-interior contact. The greater the proportion of head injuries due to upper-interior contact, the higher the potential for FMVSS No. 201 to reduce head injuries. Tables 1-1 and 1-1U showed that, overall, 42 percent of the AIS 3-to-6 head injuries and 43 percent of the AIS 4-to-6 head injuries were due to upper-interior contact immediately before FMVSS No. 201 (both 41% in the unweighted data). Table 1-4 computes those percentages for various subgroups of injuries, crashes, vehicles, and occupants. It is remarkable how little variation there is in that statistics, how close they are to 41 to 43 percent in the various subgroups. On Table 1-4, the few groups where less than 35 percent of the head injuries are due to upper-interior contact are printed in red; more than 50 percent, blue.

Occupants restrained by belts (or safety seats) have a relatively higher percentage of head injuries due to upper-interior contact (46%) than unrestrained occupants (39%). To the extent that belts reduce injuries from the steering assembly or ejection, the remaining share due to the upper interior is somewhat higher.

Life-threatening (AIS 4 to 6) and non-threatening (AIS 3) head injuries have nearly the same proportion due to the upper interior. So do the various types of injuries, except facial fractures (lower). There is little variation by crash type, except for a slightly higher proportion in subsequent-event rollovers. Vehicle type makes little difference, except for fewer upper-interior contacts in vans. Children up to 12 years old have the lowest proportion of upper-interior contacts in Table 1-4; most of the children are restrained, and too small to reach most of the upper-interior surfaces; most of the children ride in the back seat, and that probably explains the low proportion for back-seat passengers. Aside from children, there is little difference by occupant age or gender.

A clearer picture emerges in Table 1-5, limited to the head injuries of **restrained** occupants. Overall, 46 percent of the head injuries of restrained occupants are due to the upper interior. Table 1-5 uses red for proportions under 35 percent and blue for over 55 percent. As in Table 1-4, the injury severity and injury type makes little difference (except facial fractures). But impact type matters. Because belts greatly reduce the risk of ejection in rollovers, approximately 80 percent of the head injuries to belted occupants are from the upper interior – the upper roof or the roof side rail in the vast majority of cases. By contrast, the proportion is low in frontals (32 to 35%), because belts are fairly effective in keeping occupants away from the A-pillar. The proportion of upper-interior contacts is high in SUVs (56 to 58%) because they had relatively the highest proportion of rollover crashes during the model years covered in the database. Restrained children, understandably, have an especially low proportion (17 to 19%) of head injuries from contact with the upper interior.

In summary, just before FMVSS No. 201, upper-interior surfaces were responsible for consistently just over 40 percent of serious and life-threatening **head** injuries. They accounted for approximately 14 percent of **all** AIS 3-to-6 injuries and 22 percent of all life-threatening (AIS 4 to 6) injuries.

TABLE 1-4

PERCENT OF AIS 3-TO-6 HEAD INJURIES ATTRIBUTED TO UPPER-INTERIOR CONTACT
(Before FMVSS No. 201 – Unweighted 1995-2009 CDS)

	AIS 3-to-6			AIS 4-to-6		
	N Injured	Due to Upper-Interior n	%	N Injured	Due to Upper-Interior n	%
OVERALL	10,256	4,232	41	4,879	1,995	41
RESTRAINT USE						
Unrestrained	5,721	2,225	39	2,701	1,043	39
Belted (or safety seat)	4,142	1,908	46	1,997	904	45
INJURY SEVERITY						
AIS 4-6 (life-threatening)	4,879	1,995	41			
AIS 3 (not life-threatening)	5,377	2,237	42			
INJURY TYPE						
Concussion	757	346	46	474	202	43
Brain contusion	1,266	536	42	301	116	39
Unknown brain injury	5,655	2,388	42	2,880	1,210	42
Skull fracture	1,547	614	40	550	227	41
Facial fractures	341	100	29	none		
All other head injuries	690	248	36	672	240	36
IMPACT TYPE						
Frontal	2,290	858	37	1,049	376	36
Side impact	5,267	2,226	42	2,519	1,055	42
Rollover first event	1,126	461	41	529	207	39
Rollover, subsequent worst event	865	436	50	451	230	51
Rear impact or other type	443	192	43	204	99	49
VEHICLE TYPE						
Passenger car	7,035	2,941	42	3,368	1,412	42
LTV	3,221	1,291	40	1,511	583	39
Pickup truck	1,074	444	41	505	204	40
SUV	1,552	654	42	719	293	41
Van	595	193	32	287	86	30
SEAT POSITION						
Driver	6,832	2,899	42	3,282	1,376	42
Front-seat passenger	1,862	805	43	861	371	43
Back-seat passenger	1,562	528	34	736	248	34
AGE						
0-12	648	148	23	306	69	23
13-54	8,269	3,463	42	3,990	1,659	42
55+	1,336	621	46	581	267	46
GENDER						
Male	6,691	2,783	42	3,227	1,347	42
Female	3,517	1,436	41	1,629	643	39

TABLE 1-5

RESTRAINED OCCUPANTS' PERCENT OF AIS 3-to-6 HEAD INJURIES
ATTRIBUTED TO UPPER-INTERIOR CONTACT
(Before FMVSS No. 201 – Unweighted 1995-2009 CDS)

	AIS 3-to-6			AIS 4-to-6		
	N Injured	Due to Upper-Interior n	%	N Injured	Due to Upper-Interior n	%
OVERALL	4,142	1,908	46	1,997	904	45
INJURY SEVERITY						
AIS 4-6 (life-threatening)	1,997	904	45			
AIS 3 (not life-threatening)	2,145	1,004	47			
INJURY TYPE						
Concussion	295	142	48	188	82	44
Brain contusion	514	234	46	123	50	41
Unknown brain injury	2,424	1,098	45	1,256	552	44
Skull fracture	551	276	50	203	106	52
Facial fractures	124	39	31	none		
All other head injuries	234	119	51	224	114	51
IMPACT TYPE						
Frontal	893	310	35	412	131	32
Side impact	2,410	1,043	43	1,183	501	42
Rollover first event	274	215	78	130	97	75
Rollover, subsequent worst event	259	212	82	131	111	85
Rear impact or other type	234	103	44	107	50	47
VEHICLE TYPE						
Passenger car	3,144	1,384	44	1,536	669	44
LTV	998	524	53	461	235	51
Pickup truck	288	138	48	129	63	49
SUV	507	295	58	236	131	56
Van	203	91	45	96	41	43
SEAT POSITION						
Driver	2,818	1,374	49	1,360	663	49
Front-seat passenger	824	407	49	376	180	48
Back-seat passenger	500	127	25	261	61	23
AGE						
0-12	375	72	19	179	31	17
13-54	2,958	1,434	48	1,468	702	48
55+	809	402	50	350	171	49
GENDER						
Male	2,372	1,097	48	1,162	548	47
Female	1,750	801	46	825	352	43

1.2 The head-injury protection upgrade for FMVSS No. 201

FMVSS No. 201 – Occupant Protection in Interior Impact – "specifies requirements to afford impact protection for occupants."[6] Over the years, FMVSS No. 201 has primarily consisted of performance requirements limiting the amount of resistive force allowed when a headform is impacted into various sections of the vehicle interior that are typically contacted by occupants' heads during crashes.

FMVSS No. 201 was one of NHTSA's initial safety standards, effective for passenger cars on January 1, 1968, and LTVs with GVWR less than 10,000 pounds on September 1, 1981. The standard originally incorporated the Society of Automotive Engineers' 15-mph headform impact test, applying it to components identified, from the limited data available at that time, as likely head-contact areas: the top of the instrument panel, seat backs, sun visors, armrests, and other projections in head impact areas. The final rule allowed a peak deceleration of the headform of 80 g's over 3 milliseconds on the impact tests. Most cars were apparently meeting the various head-impact requirements of FMVSS No. 201 well before 1968.[7]

Analyses of improved crash data such as CDS (similar to the analyses in the preceding section) clarified that head impacts with the upper interior of vehicles continued to be the leading cause of fatal head injury for non-ejected occupants, despite the original FMVSS No. 201. Moreover, the injuries involved components not covered by FMVSS No. 201, such as pillars, roof side rails and headers, and the roof itself – and many of the injuries would not be mitigated by frontal air bags or increased use of safety belts. Based on 1988-1993 CDS data, NHTSA's 1995 *Final Economic Assessment* estimated that these impacts resulted in 2,170 fatalities and 3,630 serious non-fatal injuries per year to occupants of passenger cars and LTVs.[8]

On August 14, 1995, NHTSA issued a Final Rule extending the head injury protection requirements of FMVSS No. 201 to new target areas in order "to provide protection when an occupant's head strikes upper interior components." The existing requirements of FMVSS No. 201 remain for the original target areas. However, the new target areas in the vehicle's upper interior include the A-, B-, and other pillars, the front and rear roof header, the roof side rails, and the upper roof, among others. The speed for the free-motion headform impact test for the new areas is 24 km/h (15 mph, as in the original FMVSS No. 201) but for these targets, the HIC may not exceed 1,000 for any 36-millisecond period. Impacts may be directed from a range of vertical and horizontal angles.[9]

NHTSA believed the upgraded regulation would reduce head injuries, but did not claim it would also be effective in reducing neck injuries. The preamble of the Final Rule stated: (1) the FMH

---

[6] *Code of Federal Regulations*, Title 49, Part 571.201.
[7] Kahane, C. J. (1988, January). *An evaluation of occupant protection in interior impact for unrestrained front seat occupants of cars and light trucks*. (Report No. DOT HS 807 203). Washington, DC: National Highway Traffic Safety Administration, pp. 2-3.Available at www-nrd.nhtsa.dot.gov/Pubs/807203.PDF; Campbell, B. J (1963). *A study of injuries related to padding on instrument panels*. TRID Accession No. 00427812, Report No. VJ-1823-R2. Buffalo: Cornell Aeronautical Laboratory, 1963; Fed. Reg. 31 (December 3, 1966): 15212; SAE (1967). *1967 SAE Handbook*. New York: Society of Automotive Engineers, 1967, pp. 881-884.
[8] NHTSA (1995). *Final Economic Assessment, FMVSS No. 201, Upper Interior Head Protection*. NHTSA Docket No. 92-28-N04. Washington, DC: National Highway Traffic Safety Administration, p. I-1.
[9] *Federal Register* 60 (August 18, 1995): 43031.

was not designed to measure neck injury risk; (2) there was little evidence that adding an acceleration limit to the HIC requirement would have benefits against neck injuries; (3) no tool was available at the time to adequately model neck injury in a headform impact, but that NHTSA would consider amending the FMVSS if such a tool became available; and (4) there was no evidence that padding for head impact protection could increase neck injury risk.[10]

Manufacturers were offered a choice of several alternative phase-in schedules from September 1, 1998, to September 1, 2002. For example, they could certify the new requirements on at least 10 percent of cars and LTVs (with GVWR < 10,000 pounds) manufactured from September 1, 1998, through August 31, 1999; at least 25 percent of cars and LTVs manufactured from September 1, 1999, through August 31, 2000; at least 40 percent of cars and LTVs manufactured from September 1, 2000, through August 31, 2001; at least 70 percent of cars and LTVs manufactured from September 1, 2001, through August 31, 2002; and all cars and LTVs manufactured on or after September 1, 2002.[11]

Head-protection air bags were not yet available in production vehicles when the FMVSS No. 201 Final Rule was issued in 1995. But their development was well underway and they held great promise to further improve head-injury protection. On July 29, 1998, soon after these air bags first became available on some cars, NHTSA amended FMVSS No. 201 to facilitate their introduction on other vehicles. Recognizing that the 24 km/h (15 mph) headform test might be a problem in target areas where the undeployed air bag is stored (and, furthermore, an inappropriate test if the bag usually deploys at that speed), NHTSA offered an alternative compliance procedure. Manufacturers have the option to reduce the speed of the headform test to 19.2 km/h (12 mph) on target areas where the bag is stored, provided they can meet a 28.8 km/h (18 mph) lateral (90-degree) crash test for the full vehicle into a pole – with HIC < 1,000. The pole test simulates a head impact with the deployed bag.[12]

In summary, the principal differences between the 1995 and 1968 versions of FMVSS No. 201 are:

- Additional target areas: pillars, roof side rails and headers, and the roof itself.
- For these additional target areas, HIC may not exceed 1,000 for any 36-millisecond period (as opposed to a limit of 80 g's for 3 milliseconds on the original target areas). Impacts may be directed from a range of vertical and, in some cases, horizontal angles.

1.3 In what ways were vehicles modified?

In practice, manufacturers certified to FMVSS No. 201 by:

- Adding energy-absorbing materials such as padding, ribbing, or an "egg-crate" honeycomb configuration around target areas, or using a thicker roof liner.

---

[10] *Ibid.*
[11] *Ibid.*
[12] *Federal Register* 63 (August 4, 1998): 41451

- Adding head-protection air bags; in fact, as mentioned above, NHTSA's 1998 amendment of FMVSS No. 201 facilitated the use of air bags. Head-protection air bags include head curtains, inflatable tubular structures and torso/head combination bags.

- A combination of both, relying on energy-absorbing materials in target areas not covered by the air bag.

- Little or no change at a target area if the pre-standard vehicle would easily have met FMVSS No. 201 at that target area.

Head-protection air bags were first offered on some cars in model year 1998. In 2003, the model year completing the phase-in of the FMVSS No. 201 upgrade, 20 percent of new cars and LTVs were equipped with head-protection air bags, whereas 80 percent employed only energy-absorbing materials to meet the standard.

But on May 17, 2004, the agency issued a Notice of Proposed Rulemaking to amend FMVSS No. 214 – Side Impact Protection, proposing to add a 20-mph side impact with a pole, at a 75-degree angle (i.e., 15 degrees forward of a purely lateral impact).[13] Even though the proposal was technically not a part of FMVSS No. 201, NHTSA anticipated at the time and continues to believe that it will lead to the installation of head-protection air bags in all new vehicles, because this technology appears to be the choice to meet the pole test. A Final Rule was issued on September 11, 2007, with an amendment on June 9, 2008, to schedule the phase-in of the new requirement. The phase-in began on September 1, 2010, and it will be in effect for all new cars and LTVs after August 31, 2014.[14] In other words, the current situation, where some new vehicles have head-protection air bags and others do not, is likely to end on or before September 1, 2014. In fact, voluntary installations of head-protection air bags greatly increased in the years immediately after the 2004 NPRM. By model year 2007, well before the phase-in period, 66 percent of new cars and LTVs were already equipped with head-protection air bags, mostly curtain bags. The specific subject of this evaluation, the injury-reducing effectiveness of energy-absorbing materials **without** head-protection air bags, is already fading into the sunset.

More generally, though, the subject will still be of interest because energy-absorbing materials will continue to be needed in target areas not fully covered by the air bags – such as the upper roof, the front/rear headers and possibly the A-pillars – and may also be useful in other target areas for less severe impacts that do not deploy air bags. In fact, the only target areas where energy-absorbing materials might need to be modified are the locations that actually house the air bags. Even more generally, the evaluation is of interest because it investigates whether a performance standard based on laboratory testing with manikins and a measure of acceleration over time (HIC) translates into reductions of fatalities and injuries for human beings involved in actual crashes.

Now, let us take a closer look at the energy-absorbing materials actually used in FMVSS No. 201 vehicles not equipped with head-protection air bags.[15]

---

[13] *Federal Register* 68 (May 17, 2004): 27990.
[14] *Federal Register* 72 (September 11, 2007): 51908; *Federal Register* 73 (June 9, 2008): 32483.
[15] See also Kahane & Tarbet (2006), pp. 7-13.

In 2003, a NHTSA contractor performed a cost teardown study of the manufacturing and consumer costs of the changes made by the automotive industry to meet the FMVSS No. 201 upgrade without head-protection air bags.[16] In addition to providing cost estimates, the study describes in detail what energy-absorbing materials were actually added or modified. The contractor studied pre-standard passenger vehicles of 10 makes and models and post-standard vehicles of the same or corresponding models. The vehicles comprise a variety of manufacturers and included 6 passenger cars, a pickup truck, an SUV and 2 minivans. In addition to the contractor's own examinations of the components, the contractor received detailed information from the manufacturers, such as parts lists, identifying how vehicles were modified.

Approaches used to meet the standard include composite plastic foam padding, injection-molded ribs in parallel or egg-crate-like configurations (see Figures 1-1 and 1-2), ridges molded from composite plastic materials (Figure 1-3), crushable O-Flex tubes (Figure 1-4), stretchable fabric materials, and configuration changes of the outer trim parts to allow for the flexing of the part under load to absorb part of the impact energy.

The most popular approaches used by the manufacturers were foam padding and internal collapsible ribs. Other than configuration changes, the least expensive approach was foam padding, which is cut from pre-formed rolls of material and glued in place.

Ridges were molded into some of the foam padding. A cross-section of these ridges presents a pyramid shape with a rounded top. The energy absorption curve generated by foam padding is a straight line starting at zero (first indication of impact load) to a level of the impact load at the point of total collapse. This approach is approximately 50 percent as efficient as the square wave generated by the rib configuration.

The injection-molded ribs in the egg-crate or parallel configuration are the most expensive; however, they are also the most efficient use of material. The foam part returns to its original configuration after the impact. The ribs are thin-walled panels with parallel sides molded with the wall configuration in line with the direction of the expected impact load. The energy absorption curve generated by the collapse of the rib under impact load approaches a square wave (which is the most efficient energy absorption method). The egg-crate rib configuration is the most efficient load absorption design because the ribs are connected at 90° angles, reinforcing the load resistance capability. Parallel ribs do not have the egg-crate reinforcement feature.

Analysis of the cost teardown indicates that manufacturers redesigned their pillar trim components and headliners to comply with the standard on seven of the 10 makes and models. The remaining three models used a combination of redesign, added padding, and ribs for their post-standard vehicles. There were no changes to the C-pillar and D-rings except in the Ford vehicles. All internal ribs are made of a collapsible plastic composite material in an egg-crate (honeycomb) and/or parallel configuration. The following paragraphs describe the different approaches used for the 10 vehicles evaluated during this project.

---

[16] Ludtke, N. F., Osen, W., Gladstone, R., & Lieberman, W. (2003). *Perform cost and weight analysis, non air Bag head protection systems, FMVSS 201.* (Report No. DOT HS 809 810). Washington, DC: National Highway Traffic Safety Administration.

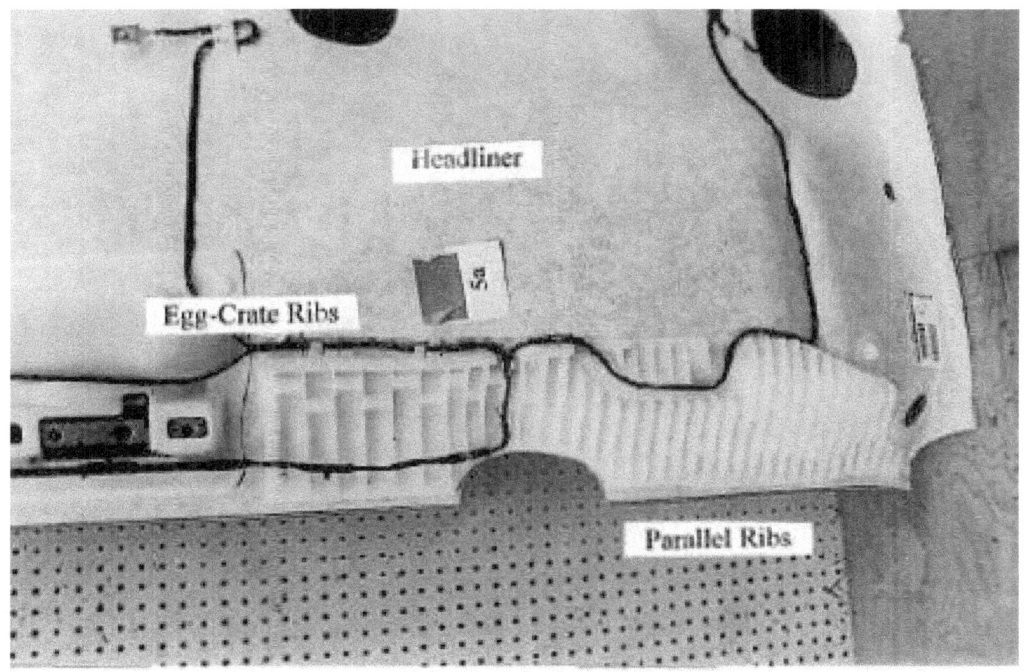

Figure 1-1: Egg-Crate and Parallel Ribs Built into the Headliner[17]

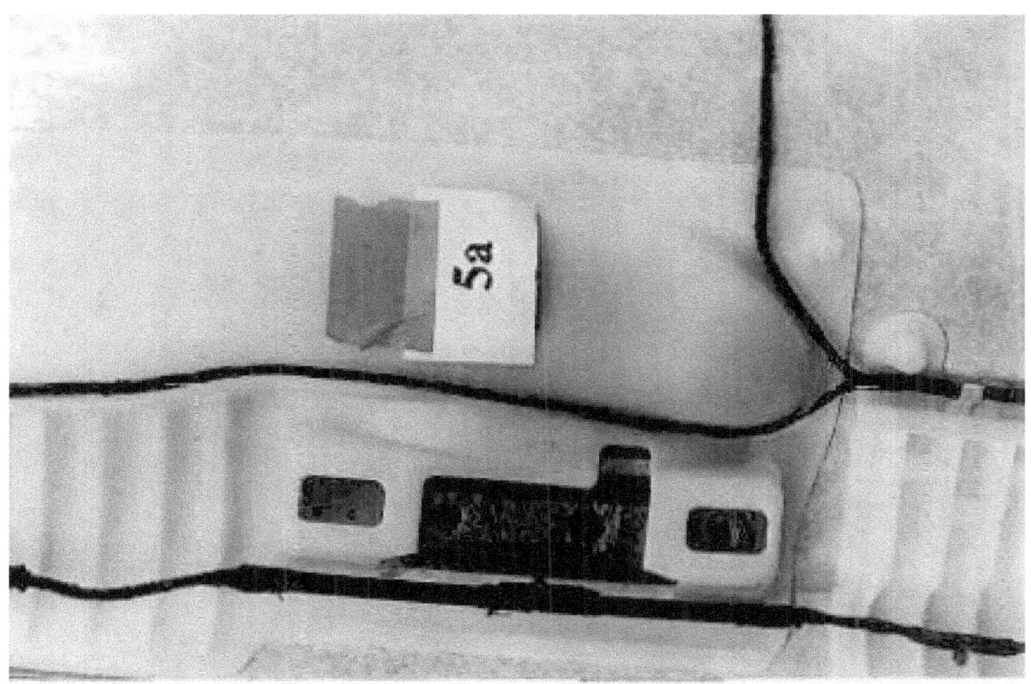

Figure 1-2: Close-Up of Parallel and Egg-Crate Ribs

---

[17] *Ibid.*, p. 2-6

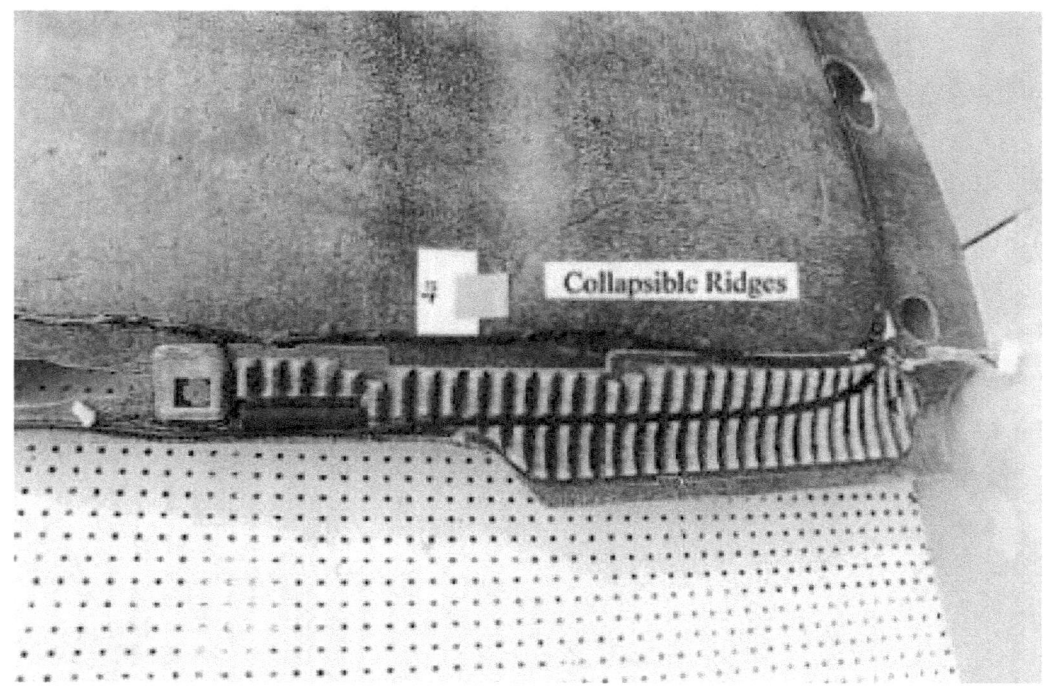

Figure 1-3: Collapsible Ridges[18]

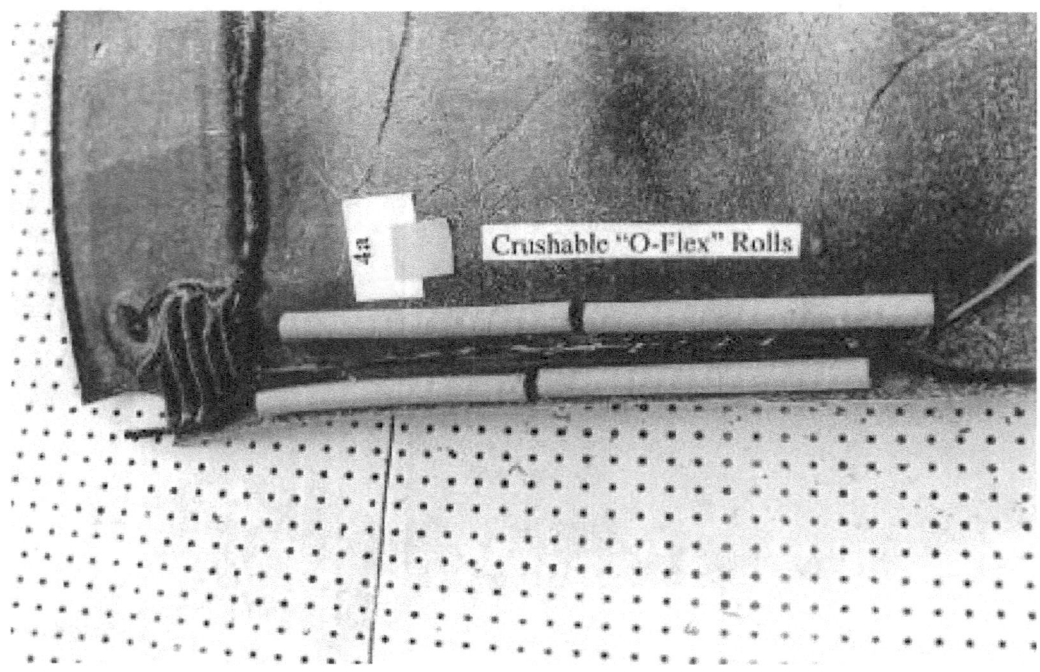

Figure 1-4: Crushable O-Flex Rolls[18]

---

[18] *Ibid.*, p. 2-7.

Dodge Caravan. The A-pillar and trim components, B-pillar and trim components, and headliner have been redesigned to comply with the standard. The cross-section of the A-pillar has been made deeper and ribs, foam strips, and a ribbed insert have been added. The B-pillar has a slightly different shape (including a larger flare where the pillar meets the roof), a loop to retain the seat belt, and two added foam pads inside the lower end of the pillar. The headliner – the interior lining of the roof – has six additional plastic panels (combination parallel and egg-crate configuration) across the front and down both sides. Three of these panels serve as both air ventilation ducts and energy absorption devices. The other three panels are devoted to energy absorption purposes. Two small Styrofoam pads were added to the headliner near the top of the B-pillar.

Ford Crown Victoria. The upper interior components have been redesigned to comply with the standard. The A-, B-, and C-pillar trim has been changed and internal parallel ribs and fasteners have been added. The D-ring cover material has been changed to include internal ribs. The grab handles and hooks have different energy absorbing materials and collapsible fasteners. The headliner material has been changed and foam padding added. O-Flex crushable tubes and egg-crate pieces have been added around the side edges.

Ford F-150 pickup truck. The upper interior components have been redesigned to comply with the standard. The A- and B- pillar trim has been changed and internal parallel ribs and fasteners have been added. In addition, an O-Flex crushable tube has been added to the B-pillar trim. The D-ring covers are made of a different material that includes internal collapsible ribs. The grab handle material has been changed to an energy absorbing type and is supported by a bracket attached to the inner A-pillar structure. Blocks of foam padding and O-Flex crushable tubes have been added to selected areas around the outer edge of the headliner, and a rib cartridge has been added to the foam at the side rail near the A-pillar. The overhead console is made of a new material that has better performance at impact.

Ford Taurus. The upper interior components have been redesigned to comply with the standard. The shape of the A-pillar has been changed. Foam padding inside the trim and collapsible fasteners has been added to the A-pillar trim. A metal strap and washer protect the foam padding from a sheet metal flange. The B-pillar trim material and shape has been changed. Internal parallel ribs and fasteners have been added to the C-pillar. The D-ring covers are made of a different material that includes internal collapsible ribs. The headliner is made of a different material that is slightly thicker than that of the pre-standard model, and three blocks of foam have been placed adjacent to the side rails. These blocks are about six inches in width and the combined length extends from the front to the rear of the headliner.

Honda Accord. The headliner has been redesigned and padding has been added to the A-pillar trim to comply with the standard. The padding is made from a composite plastic material. Collapsible vertical plastic ridges have been added to the outer edges on the rear of the headliner, while plastic O-Flex crushable tubes have been added to the outer edges on the front. There has been no change to the B-pillar.

Jeep Grand Cherokee. The A-pillar, B-pillar, and headliner have been redesigned to comply with the standard. The cross-section of the A-pillar has been widened (3½ inches versus 2 inches). An extra 12 internal egg-crate ribs, with a strip of foam along the inside edge of the pillar, have been added. The B-pillar also has a wider cross-section (about ½ inch) and an additional 34 internal parallel ribs. The post-standard headliner has nine added ribbed panels, made of molded plastic collapsible foam with ridged features, which are glued to the roof side of the panel.

Kia Spectra. Padding has been added to the A-pillar trim, B-pillar trim, and headliner to comply with the standard. A Styrofoam pad is glued inside the A- and B-pillar trim. Five foam pads have been added to the left and right side of the headliner and glued to the inside. The pads are not a molded shape but simply rectangular blocks sheared from a one-inch thick sheet of foam.

Pontiac Montana. The A-pillar, B-pillar, and headliner have been redesigned to comply with the standard. The A-pillar trim panel cross-section shape is ½ inch wider with small thickness changes. There are 26 additional internal parallel ribs, and a triangular foam pad has been added near the base of the panel. The B-pillar has an additional 17 internal parallel ribs, a larger cross-section or "footprint" where the B-pillar meets the roof, and an enlarged upper attaching point that has an additional seven internal parallel ribs. The headliner has an additional six plastic risers, which look like heavy duty "bubble wrap" and hold the foam pads in place. The center seat head protection pads are Styrofoam that are glued directly to the headliner

Toyota Camry. The upper interior components have been redesigned to comply with the standard. The A-pillar trim has been changed to include internal egg-crate ribs, while the B-pillar trim has been changed to include internal parallel ribs and reinforcement padding. Changes have been made in the shapes of the rear header, the roof side rails, and the roof upper header. Plus, reinforcements have been added to the roof side rails to improve their head protection capability.

Volkswagen Jetta. The upper interior components have been redesigned to comply with the standard. The post-standard A-pillar extends from the dash to the roof, whereas the pre-standard pillar had extended from the floor to the roof. The cross-section of the A-pillar trim panel has been enlarged and is covered with a stretchable fabric over a 1/32-inch thick felt. The shape of the cross-section of the B-pillar has also been changed and two Styrofoam strips have been added. The trim panel is covered with the same stretchable fabric as the A-pillar. The headliner has been redesigned to make it wide enough to reach around the roof outer perimeter structure to afford better head protection. The increase in material for the headliner is nearly 16 percent.

Fundamentally,

- The A-pillar was substantially modified in every vehicle. Generally, though, it is not the steel structures of pillars that were modified, but rather their interior lining or "trim panels" – the material between the pillars and the occupant compartment.

- The B-pillar and headliner (interior lining of the roof) were modified in almost every vehicle.

- According to the preceding descriptions, other pillars, roof side rails, the front header and the rear header remained largely unchanged in these specimen models (but the HIC tests described in the next section actually showed a significant improvement after FMVSS No. 201, in these models, in impacts with the roof side rails).

NHTSA has little information about what happened to these energy-absorbing materials after manufacturers began to offer head-protection air bags (other than an intuitive feel that many of the materials would be unchanged). NHTSA's cost analyses of FMVSS No. 201 comprise 15 makes and models including 8 that initially certified to FMVSS No. 201 with energy-absorbing materials only and subsequently offered standard or optional head-protection air bags one or more years later – i.e., they upgraded head impact protection in two separate stages.[19] The contractor studied the price of parts to replace pillars in these 8 models before and after head-protection air bags became available. But in 5 of the 8 models, it appears the contractor's "before" vehicles were actually pre-FMVSS No. 201. The appropriate comparison – FMVSS No. 201-certified without air bags versus FMVSS No. 201-certified with air bags – was only pursued on 3 of the 8 models. For those three, the average cost of the replacement parts was slightly higher in the vehicles with the air bags. That suggests energy-absorbing materials were not degraded, at least on these three models, after air bags became available.[20]

1.4   HIC test results before and after FMVSS No. 201

For an initial study of the efficacy of FMVSS No. 201, NHTSA compared the Head Injury Criterion on the headform impact test, HIC(d) at matching target locations in the same or similar makes and models before and after FMVSS No. 201.[21] The "after" test results are from NHTSA compliance tests for FMVSS No. 201. From the 68 vehicles that had been tested as of 2003, NHTSA purposively selected 15 high-sales models of passenger cars, pickup trucks, SUVs and minivans. In 2004, a contractor purchased 15 pre-standard vehicles of the same makes and models and performed identical headform impact tests, at the same locations, as in the compliance tests. (When an exact match was impossible because a model was discontinued or its interior redesigned, the closest corresponding model or location was selected.) In all, there were 154 matched pairs of impact tests in pre- and post-standard vehicles. The 15 models and their corresponding pre-standard vehicles were:

---

[19] Ludtke et al. (2003); Ludtke, N. F., Osen, W., Gladstone, R., & Lieberman, W (2004). *Perform Cost and Weight Analysis, Head Protection Air Bag Systems, FMVSS 201*. (Report No. DOT HS 809 842). Washington, DC: National Highway Traffic Safety Administration.
[20] Ludtke et al. (2004), pp. 3-47 – 3-54 and Appendix A; the 8 models in the study that initially certified without air bags and later received air bags are Jeep Grand Cherokee, Ford Taurus, Ford Crown Victoria, Ford F-150, Pontiac Montana, Honda Accord, Toyota Camry and Hyundai Accent; however, it appears from the report that the specimen without air bags was 201-certified only on the Grand Cherokee, the Camry and the Accent.
[21] Kahane & Tarbet (2006).

	Post-Standard Vehicle		Pre-Standard Vehicle
1999	Jeep Grand Cherokee	1996	same make and model
2000	Dodge Neon 4-door	1997	same make and model
2002	Dodge Grand Caravan	1996	same make and model
2002	Ford F-150 supercab	1998	same make and model
2002	Ford Explorer 4-door	1998	same make and model
1999	Ford Windstar	1998	same make and model
2001	Buick LeSabre 4-door	1998	same make and model
1999	Chevrolet Silverado x-cab	1998	Chevrolet C-1500 x-cab
2002	Chevrolet Trailblazer 4-door	1996	Chevrolet Blazer 4-door
2001	Nissan Sentra 4-door	1998	same make and model
1999 & 2003	Honda Accord 4-door	1996	same make and model
2003	Toyota Corolla 4-door	1998	same make and model
1999 & 2002	Toyota Camry 4-door	1998	same make and model
2003	Toyota Tacoma Xtracab	1997	same make and model
2002	Kia Spectra 4-door	1997	Kia Sephia 4-door

Table 1-6 shows that HIC(d) averaged 909.9 in 154 head impact tests on pre-standard vehicles, ranging from as low as 426 to as high as 1,767. In compliance tests of post-standard vehicles, the 154 impacts to matching locations in the same makes and models resulted in a range of HIC(d) from 373 to 986 and an average of 667.5. That is an average improvement of 242.4 units of HIC per test.

TABLE 1-6

AVERAGE HIC(d) BEFORE AND AFTER FMVSS No. 201
(154 impact locations in 15 makes and models)

	Average	Lowest	Highest	Standard Error	t-test
Pre-standard HIC(d)	909.9	426	1,767		
Post-standard HIC(d)	667.5	373	986		
HIC(d) improvement	242.4	− 420	1,194	24.1	10.06

On the 154 matched pairs of impacts, the improvement in HIC(d) ranged from –420 (i.e., it became worse) to +1194, the average improvement being 242.4. The standard error of the improvement is 24.1. Because t = 242.4/24.1 = 10.06 is much more than the 95th or even the 99th percentile of a t distribution with 153 degrees of freedom, it is a statistically significant improvement.

Furthermore, HIC(d) exceeded 1,000 in 47 of the 154 locations tested in pre-standard vehicles, but was less than 1,000 in each of the 154 locations in the post-standard vehicles. Only 2 of the

15 pre-standard vehicles, but all of the post-standard vehicles had HIC(d) ≤ 1,000 at each location.

HIC(d) was reduced by an average of 476 units in impacts with the A-pillar, 61 in impacts with the B-pillar, 105 with other pillars, 279 with the upper roof and 245 with the roof side rail. All of these improvements were statistically significant. There was a non-significant average increase of HIC(d) by 35 in impacts with the front or rear header. Performance improvement was the largest on the A-pillar, where pre-standard performance was the poorest. In the pre-standard vehicles, HIC(d) averaged 1154 for the A-pillar, 737 for the B-pillar, 747 for other pillars, 930 for the roof side rail, 941 for upper roof and 543 for the front and rear header. After FMVSS No. 201, HIC(d) averaged 679 for the A-pillar, 676 for the B-pillar, 642 for other pillars, 684 for the roof side rail, 662 for upper roof and 578 for the front and rear header. In other words, post-standard HIC(d) was nearly uniform at about 650 across the upper interior, with larger improvements on the components where HIC(d) was originally higher.[22]

Overall, HIC(d) improved by an average of 172 in the tests of passenger cars, and by 201 in LTVs. Both reductions are statistically significant.

1.5 Fatality reduction by head-protection air bags in side impacts

In 2007, NHTSA published initial statistical analyses of the fatality-reducing effectiveness of head-protection air bags in crashes, including head curtains, inflatable tubes, and head-torso combination bags.[23] Because the first generation of these air bags was designed to deploy only in side impacts, the analyses only address side impacts; however, they distinguish between nearside occupants – e.g., the driver in a left-side impact – and far-side occupants – e.g., the driver in a right-side impact. (Starting in mid-model year 2002, Ford began to offer head curtains on selected vehicles that could deploy and remain inflated for some time in rollover crashes.) Because most of the early air bags were in passenger cars, not LTVs, the analyses primarily address passenger cars.

Table 1-7, based on FARS and GES data, computes fatality rates for drivers and right-front passengers of passenger cars who are nearside occupants in side impact crashes. It compares the fatality rate per 1,000 occupants with standard head-protection air bags plus torso bags to the rate in the same makes and models without side air bags.

---
[22] Kahane & Tarbet (2006), pp. 19-21.
[23] Kahane (2007).

TABLE 1-7[24]

CAR MODELS THAT ONCE HAD NO SIDE AIR BAGS
AND NOW HAVE STANDARD TORSO BAGS PLUS HEAD PROTECTION
FATALITIES PER 1,000 NEARSIDE FRONT-SEAT OCCUPANTS IN SIDE IMPACTS
(1993-2005 FARS and GES data; fatalities adjusted upward for safety belt use;
all cars are 214-certified, equipped with dual frontal air bags, and model year 1994-2003)

	Nearside Fatalities	Nearside Occupants	Fatality Rate	Fatality Reduction
Without side air bags	660	163,018	4.05	
With torso bags + head protection	265	95,349	2.78	31 %

When these models did not have side air bags, the fatality rate was 4.05 per 1,000 nearside occupants. When the same models were equipped with torso plus head air bags, the fatality rate was 2.78. That is a statistically significant 31-percent reduction from the rate without side air bags.

Table 1-8 is another analysis of the same makes and models, but based only on FARS data. It compares the ratio of nearside to longitudinal (impact location 12:00 or 6:00 – purely frontal or purely rear impacts) fatalities with standard torso bags plus head protection to the ratio in the same models without side air bags. The longitudinal impacts are a control group because the head-protection air bags do not deploy. When the cars did not have side air bags, the fatality risk ratio was .531. With torso bags plus head protection, the fatality risk ratio was .424. That is a 20-percent reduction. It is not statistically significant.

TABLE 1-8[25]

CAR MODELS THAT ONCE HAD NO SIDE AIR BAGS
AND NOW HAVE STANDARD TORSO BAGS PLUS HEAD PROTECTION
NEARSIDE VERSUS LONGITUDINAL FATALITIES
(1993-2005 FARS; fatalities adjusted upward for safety belt use;
all cars are 214-certified, equipped with dual frontal air bags, and model year 1994-2003)

	Nearside Fatalities	12:00 or 6:00 Fatalities	Risk Ratio	Nearside Reduction
Without side air bags	746	1,404	.531	
With torso + head protection	309	728	.424	20 %

---

[24] *Ibid.*, p. 98.
[25] *Ibid.*, p. 99.

Table 1-9 expands the analysis of Table 1-8 to also include models of passenger cars that offered a choice between no bags and torso bags plus head protection. With the additional data, the fatality reduction is a statistically significant 19 percent.

TABLE 1-9[26]

CAR MODELS THAT ONCE HAD NO SIDE AIR BAGS
AND NOW HAVE STANDARD TORSO BAGS PLUS HEAD PROTECTION,
OR OFFERING A CHOICE BETWEEN NO AIR BAGS AND TORSO + HEAD AIR BAGS
NEARSIDE VERSUS LONGITUDINAL FATALITIES
(1993-2005 FARS; fatalities adjusted upward for safety belt use;
all cars are 214-certified, equipped with dual frontal air bags, and model year 1994-2003)

	Nearside Fatalities	12:00 or 6:00 Fatalities	Risk Ratio	Nearside Reduction
Without side air bags	1,051	2,140	.491	
With torso + head protection	433	1,092	.396	19 %

The average of those three results is **24 percent**.[27] The 90-percent confidence bounds for that estimate, taking into account not only sampling error but computational uncertainty (varied results when different analyses are applied to fundamentally the same data) range from **4 to 42 percent**.[28] Analyses of the limited data for LTVs suggest that effectiveness may be about the same as for cars in nearside impacts.[29]

Statistical analyses of crash data also show significant reductions of fatality risk for head curtains or inflatable tubes (but not for head-torso combination bags) in **far-side** impacts to passenger cars (but not LTVs). Specific mechanisms whereby side air bags mitigate injuries in far-side impacts have not yet been widely demonstrated or quantified by testing – but head curtains are designed to deploy in far-side impacts, cover areas responsible for a large proportion of the life-threatening injuries, and are likely to remain at least partially inflated by the time the far-side occupant contacts them. For example, Table 1-10, a FARS-GES analysis comparable to Table 1-7, shows a statistically significant 35-percent reduction in far-side fatality risk with inflatable curtains or tubes.

---

[26] *Ibid.*, p. 100.
[27] The actual effectiveness estimates (not rounded) are 31.29%, 20.18% and 19.37%.
$1 - \exp\{[\log(1 - .3129) + \log(1 - .2018) + \log(1 - .1937)] / 3\} = 1 - \exp(-.2720) = 23.81$ percent
[28] *Ibid.*, p. 123.
[29] *Ibid.*, pp. 119-120.

TABLE 1-10[30]

CAR MODELS THAT ONCE HAD <u>NO</u> SIDE AIR BAGS
AND NOW HAVE <u>STANDARD</u> HEAD CURTAINS OR INFLATABLE TUBES
FATALITIES PER 1,000 FAR-SIDE FRONT-SEAT OCCUPANTS IN SIDE IMPACTS
(1993-2005 FARS and GES data; fatalities adjusted upward for safety belt use;
all cars are 214-certified, equipped with dual frontal air bags, and model year 1994-2003)

	Far-side Fatalities	Far-side Occupants	Fatality Rate	Fatality Reduction
Without side air bags	147	69,811	2.11	
With curtains/tubes + torso bags	63	46,172	1.37	35 %

---

[30] *Ibid.*, p. 98.

CHAPTER 2

EFFECT OF FMVSS No. 201 ON AIS 3-TO-6 HEAD INJURIES FROM UPPER-INTERIOR SOURCES: ANALYSES OF 1995-2009 CDS DATA

2.0 Summary

The 1999-2003 upgrade of FMVSS No. 201 substantially reduced the Head Injury Criterion (HIC) in test impacts by a headform into upper-interior contact areas. HIC was reduced by an average of 242 units per impact. Statistical analyses of head injuries from upper-interior contact in actual crashes generally agree with the test results. AIS 4-to-6 head injuries from upper-interior contact were reduced by 32 percent; the reduction is statistically significant at the two-sided .05 level. FMVSS No. 201 is especially effective in first-event rollovers, in head impacts with the A-pillar or roof interior (where HIC was most strongly reduced in headform tests), and against concussions.

2.1 A database of AIS 3-to-6 injuries before and after FMVSS No. 201

The Crashworthiness Data System (CDS) of the National Automotive Sampling System (NASS) is a national probability sample of passenger vehicles, including cars and LTVs (pickup trucks, SUVs and vans with less than 10,000 pounds gross vehicle weight rating) involved in crashes where at least one vehicle was towed from the scene. CDS data from 1995 to 2009 are analyzed because they extend several years before and after the phase-in of FMVSS No. 201 (1999-2003) and because the definitions of injuries and injury sources did not change.

Injury data. CDS documents occupants' injuries based on data from hospitals, treatment facilities, and autopsies. The analyses consider injuries rated 3 to 6 on the AIS:

- 3 Serious (but not life-threatening)
- 4 Severe (life-threatening, survival probable)
- 5 Critical (life-threatening, survival uncertain)
- 6 Not survivable with current medical technology

An occupant may have more than one such injury.

CDS documents the body region of the injury, the lesion, and the system or organ involved. In the analyses that follow, AIS 3-to-6 injuries are classified as "head injuries" that have a chance to be mitigated by FMVSS No. 201, "control group injuries" unlikely to be affected by FMVSS No. 201, or neither. All burns (as defined by the lesion) are assigned to the control group. Except for burns, any injury with body region H (head) or F (face) is a head injury. Injuries with body regions O (whole body), U (unknown), and N (neck) are not assigned to either group, because it is unclear if FMVSS No. 201 might have an effect. They are excluded from the analyses. Injuries to all other body regions, including the chest, abdomen, back, arms, and legs are in the control group.

The injury source (contact point) is also an important factor in the analyses. The following injury source codes comprise the "upper interior" of the vehicle:

> A-pillar (injury source codes 53 and 103);
> B-pillar (54 and 104);
> Other pillar (55 and 105);
> Front header (201);
> Front header plus sun visor (20);
> Rear header (202);
> Roof side rail (203 and 204);
> Roof interior, roof maplite, sunroof (205, 206 and 207);
> Windshield plus surrounding structures (A-pillar and/or front header, 15 and 16); and
> Side window plus surrounding structures (pillars and/or roof side rail, 59 and 109).

The last two are included because they involve upper-interior structures (pillars, headers, roof side rails) even though they also involve other components (glazing). All other known injury source codes (including non-contact injuries such as code 603) are not considered to be part of the upper interior: they are part of the control group. Injuries of unknown source (codes 697 or blank) are excluded from the study.

The analysis approach is to count the number of AIS 3-to-6 head injuries from upper interior sources and compute its ratio to control group injuries: AIS 3-to-6 injuries that are not head injuries **and/or** are not from upper interior sources. The ratios are compared for pre- and post-FMVSS No. 201 vehicles.

FMVSS No. 201 certification. On August 14, 1995, NHTSA issued the final rule extending the head-injury protection requirements of FMVSS No. 201 to the new target areas of the upper interior. Manufacturers were offered a choice of several alternative phase-in schedules from September 1, 1998, to September 1, 2002 – i.e., beginning in model year 1999, with full phase-in by model year 2003.[31] Every year, the manufacturers send letters to NHTSA with information to enable the agency to prepare its annual *Buying a Safer Car* reports that describe the safety features of new vehicles. The manufacturer letters, themselves, may include confidential information, but the data presented in *Buying a Safer Car* is, of course, public information. During model years 1999 to 2002, the letters specified what vehicles had been FMVSS No. 201-certified in response to the phase-in requirements. This material is summarized in a single "Head Injury Protection" column in *Buying a Safer Car* for those years.[32]

All vehicles were FMVSS No. 201-certified in model year 2003 (i.e., after September 1, 2002); none in 1998 (i.e., before September 1, 1998). Furthermore, each of the 19 distinct makes and

---

[31] *Federal Register* 60 (August 18, 1995): 43031.
[32] NHTSA. (1998). *New Car Safety Features 1999.* (Report No. DOT HS 808 808). Washington, DC: National Highway Traffic Safety Administration; NHTSA (2000). *Buying a Safer Car 2000.* (Report No. DOT HS 809 046). Washington, DC: National Highway Traffic Safety Administration; NHTSA. (2001). *Buying a Safer Car 2001.* (Report No. DOT HS 809 152). Washington, DC: National Highway Traffic Safety Administration; NHTSA. (2002). *Buying a Safer Car 2002.* (Report No. DOT HS 809 409). Washington, DC: National Highway Traffic Safety Administration.

models studied in NHTSA's cost analysis of FMVSS No. 201[33] and/or its analysis of HIC test results before and after FMVSS No. 201[34] underwent substantial modifications to meet the head-impact upgrade. As a result, we may assume (with possible, relatively rare exceptions) that vehicles not certified to FMVSS No. 201, including all pre-1999 vehicles, probably would not have met the requirements of the head-impact upgrade – and we may call them "pre-FMVSS No. 201 vehicles."

Appendix A of this report lists every make and model that was sold in at least one of model years 1999-2002 and specifies the model year when FMVSS No. 201 certification began. All model years before that, including any pre-1999 vehicle (even of make-models not listed in Appendix A) are "pre-standard." All model years after the certification, including any 2003 and later vehicle (including make-models not listed) are "post-standard." Questionable model years (e.g., certification mid-year; certification dependent on non-VIN-identifiable features such as sliding roofs) are excluded from the analysis.

CDS analysis file. Because FMVSS No. 201 applies to many parts of the upper interior, from front to back and from left to right, AIS 3-to-6 injuries to any occupant at a designated seating position are included: drivers, front passengers and rear passengers, outboard and center (codes 11, 12, 13, 21, 22, 23, 31, 32, and 33). Excluded are occupants not seated at a specific position within the vehicle (e.g., code 28) or not really in the vehicle (e.g., in the bed of a pickup truck). Occupants may be of any age.

To maximize available data, a range of model years 1985 to 2009 is initially considered. However, a portion of the data needs to be excluded to provide at least a degree of uniformity between the pre- and post-FMVSS No. 201 cases.

Above all, this is a study of the effect of head-impact protection **without** head-protection air bags. Therefore, all outboard occupants at seats equipped with head-protection air bags are excluded from the analysis, regardless of whether these air bags are the curtain, inflatable-tube or head-torso-combination type, and regardless of whether the air bags deployed. The determination of whether a seat is equipped with a head-protection air bag (or, for that matter, any other type of air bag) is based on decoding the VIN. Since 1991, NHTSA staff has maintained a series of VIN analysis programs for use in evaluations. The programs are available to the public.[35] When a determination could not be made from the VIN, the CDS variable RESTYPE was consulted. Appendix A lists some of the make and models that were equipped with head-protection air bags upon or even before FMVSS No. 201 certification, including all or most Mercedes, Audi, BMW, Jaguar, Saab and Volvo.

Also excluded from the analysis, to help make the data more uniform, are:

---

[33] Ludtke, Osen, Gladstone, & Lieberman (2003).
[34] Kahane & Tarbet (2006).
[35] Kahane (2007), p. 69.

- Any drivers or right-front passengers at seats equipped with 2-point automatic belts, regardless of whether the belts were in use. (But 3-point automatic belts are not excluded, because they perform similar to manual 3-point belts.[36])

- Any drivers or right-front passengers in **frontal** impacts (GAD1 = F, excluding first-event rollovers and certain other non-collisions) at seats **not** equipped with frontal air bags. This excludes nearly all front-outboard occupants of pre-1990 vehicles in frontals. The reason we want the vehicles uniformly equipped with frontal air bags is that the bags substantially reduce control-group injuries: torso injuries and head injuries not due to upper-interior contact. As a result, the ratio of upper-interior head injuries to control-group injuries (our measure of risk) is higher in vehicles with air bags than without them. But this exclusion is unnecessary for non-frontal crashes, where air bags would not deploy, or for back-seat occupants, where frontal air bags do not exist.

- Any vehicle equipped with torso bags in **non-frontal** impacts (GAD1 ≠ F, plus first-event rollovers and non-collisions). The rationale here, for side impacts at seats equipped with torso bags, is analogous to frontal impacts and frontal air bags. But in the frontals, where most vehicles have been equipped with frontal air bags since the early 1990s, we excluded the vehicles without them. Here, where only a relatively small percentage of fairly new vehicles have torso bags, it is more expedient to exclude the vehicles with them (and, in fact, exclude them for all seat positions in all non-frontal impacts).

The analysis file includes 31,160 individual AIS 3-to-6 injuries of known type, severity and source, including 6,606 in FMVSS No. 201-certified vehicles.

2.2   Overall reduction of AIS 3-to-6 head injuries from upper-interior sources

Table 2-1U treats the 31,160 injuries on the CDS-based analysis file as if they were individual cases and it computes effectiveness based on the unweighted case counts. Table 2-1U shows 4,226 AIS 3-to-6 head injuries from upper-interior sources and 20,328 other AIS 3-to-6 injuries (i.e., head injuries from sources other than the upper interior, or injuries from any source to the torso, arms or legs) before FMVSS No. 201, a risk ratio of .208. In the FMVSS No. 201-certified vehicles without head-protection air bags, there were 1,067 head-upper interior injuries and 5,539 other injuries, a risk ratio of .193. That is a 7-percent reduction of head-upper-interior injuries relative to the control group.

---

[36] Kahane, C. J. (2000, December). *Fatality reduction by safety belts for front-seat occupants of cars and light trucks: Updated and expanded estimates based on 1986-99 FARS data.* (Report No. DOT HS 809 199, pp. 40-41). Washington, DC: National Highway Traffic Safety Administration. Available at www-nrd.nhtsa.dot.gov/Pubs/809199.PDF.

TABLE 2-1U

OVERALL EFFECT OF FMVSS No. 201 ON AIS **3-to-6** HEAD INJURIES
FROM UPPER-INTERIOR SOURCES, PASSENGER CARS AND LTVs
(Unweighted 1995-2009 CDS)

	Head-Upper Interior Injuries	All Other AIS 3 to 6 Injuries	Risk Ratio	Reduction (%)
Pre-FMVSS No. 201	4,226	20,328	.208	
FMVSS No. 201-certified	1,067	5,539	.193	7

CDS, however, is not merely a collection of individual cases, but a probability sample of the nation's towaway crashes. Each case has a ratio weight factor equal to the inverse of its probability of selection. For unbiased national estimates of totals, each case needs to be weighted by RATWGT. Table 2-1W is the same analysis as Table 2-1U, but its four data cells are the sum of the RATWGTs for the cases in that cell, rather than just a count of the cases. Each cell in Table 2-1W is about 50 times as large as in Table 2-1U, because the average of RATWGT for these AIS 3-to-6 injury cases is approximately 50 (i.e., CDS is close to a 2 percent sample of the nation's injuries at these severity levels). In the weighted data, the injury reduction associated with FMVSS No. 201 is 11 percent, higher than the 7 percent in Table 2-1U.

TABLE 2-1W

OVERALL EFFECT OF FMVSS No. 201 ON AIS **3-to-6** HEAD INJURIES
FROM UPPER-INTERIOR SOURCES, PASSENGER CARS AND LTVs
(Weighted 1995-2009 CDS)

	Head-Upper Interior Injuries	All Other AIS 3 to 6 Injuries	Risk Ratio	Reduction (%)
Pre-FMVSS No. 201	229,025	1,316,059	.174	
FMVSS No. 201-certified	49,517	318,437	.156	11

The weighted results are the principal estimates, being unbiased and nationally representative. The unweighted results were shown first only because they are a simpler analysis, but they are really just an intermediate step in generating the weighted results. Nevertheless, the unweighted results have one advantage, namely less sampling error, and this makes them worth considering in combination with the weighted results. Weighted data are more prone to statistical uncertainty, because a few high-RATWGT cases in one cell or another can distort the results. The unweighted data yield a more statistically precise estimate, at the expense of unknown bias. If the weighted and unweighted results clearly diverge – e.g., one is strongly positive and the other is negative – that could be an indication of sampling error in the weighted result and a caution flag about its accuracy.[37]

The injury reductions are substantially larger for the life-threatening AIS 4-to-6 injuries. Table 2-2U shows a 20-percent reduction in the unweighted data, while Table 2-2W estimates a 32 percent effect in the weighted data. Note also that a larger proportion of the AIS 4-to-6 injuries are head-upper interior than at the AIS 3 level (as evidenced by higher risk ratios than in Tables 2-1U and 2-1W).

TABLE 2-2U

OVERALL EFFECT OF FMVSS No. 201 ON AIS **4-to-6** HEAD INJURIES
FROM UPPER-INTERIOR SOURCES, PASSENGER CARS AND LTVs
(Unweighted 1995-2009 CDS)

	Head-Upper Interior Injuries	All Other AIS 4 to 6 Injuries	Risk Ratio	Reduction (%)
Pre-FMVSS No. 201	1,993	6,931	.288	
FMVSS No. 201-certified	424	1,848	.229	20

TABLE 2-2W

OVERALL EFFECT OF FMVSS No. 201 ON AIS **4-to-6** HEAD INJURIES
FROM UPPER-INTERIOR SOURCES, PASSENGER CARS AND LTVs
(Weighted 1995-2009 CDS)

	Head-Upper Interior Injuries	All Other AIS 4 to 6 Injuries	Risk Ratio	Reduction (%)
Pre-FMVSS No. 201	103,829	354,393	.293	
FMVSS No. 201-certified	18,791	93,892	.200	32

Significance testing. The SURVEYFREQ procedure of SAS analyzes contingency tables such as the four preceding ones and generates a Rao-Scott chi-square (χ^2) statistic. It has the same critical values as a conventional χ^2, namely 3.84 is its 95th percentile (the minimum amount needed for statistical significance at the two-sided .05 level), 2.71 is its 90th percentile

---

[37] Of course, if CDS is used to estimate national totals, such as the number of AIS 3 injuries in frontal crashes during 2005, only the weighted data are meaningful. Population-based injury rates, such as the number of AIS 3 injuries per 100 occupants involved in towaway crashes, are likewise meaningless with unweighted data, because in the CDS sample design, the low-injury cases (the denominator) are generally sampled at a lower rate than the high-injury cases (the numerator). Unless the cases are weighted by RATWGT, the injury rates are far too high.

(significance at the one-sided .05 level), and 1.64 is its 80th percentile. The procedure first calculates the conventional χ^2 if the cell counts had been independent observations from a simple random sample (SRS). The Rao-Scott χ^2 is the SRS χ^2 divided by the design effect (DEFF) due to the numbers actually deriving from a cluster sample of 24 primary sampling units (PSU) and, in Tables 2-1W and 2-2W, being weighted by RATWGT. CDS differs from a simple random sample in numerous ways:

- CDS crash cases are not selected all over the United States by SRS but are clustered in groups of contiguous counties called primary sampling units and sometimes further clustered within the PSUs.

- The same occupant may have more than one AIS ≥ 3 injury; thus, the injury cases are not necessarily independent observations.

- Likewise, one vehicle may have more than one injured occupant, and one crash may involve more than one vehicle with injured occupants.

- CDS cases within the PSUs do not have equal probability of selection, but different probabilities according to a stratified sampling scheme.

The Rao-Scott χ^2 statistic for Table 2-2W is 7.67. In other words, the 32-percent reduction of AIS 4-to-6 head-upper interior injuries after FMVSS No. 201 is statistically significant at the two-sided .05 level. The corresponding 20-percent reduction in the unweighted data (Table 2-2U) has χ^2 = 3.37, statistically significant at the one-sided .05 level. However, neither of the estimated reductions for AIS 3-to-6 injuries is statistically significant; the chi-squares are .89 for Table 2-1W and .45 for Table 2-1U.

2.3 Effect on AIS 4 to 6 for specific impact, vehicle, occupant and injury types

Table 2-3 estimates the post-FMVSS No. 201 reductions of AIS 4-to-6 head injuries due to upper-interior contact for various impact, vehicle, occupant, and injury types. The left half of the table estimates effectiveness for all occupants, belted and unbelted. The right half is limited to people who wear seat belts. For each group, Table 2-3 shows the percent injury reduction and the Rao-Scott χ^2 statistic. Reductions that are statistically significant at the two-sided .05 level ($\chi^2 > 3.84$) are printed in bold type and shaded in bright blue; significance at the one-sided .05 level ($\chi^2 > 2.71$) is shaded in pale blue; a blue border indicates χ^2 in the 80th to 90th percentiles (1.64-2.70). Negative estimates are printed in red (none are significant). The first two estimates reiterate Tables 2-2W and 2-2U, namely, a statistically significant overall 32-percent reduction of AIS 4-to-6 head injuries due to upper-interior contact, with a corresponding 20--percent reduction in the unweighted data.

Table 2-3: Reduction (%) of **AIS 4-to-6** Head Injuries Due to Upper Interior Contact, Relative to Pre-FMVSS No. 201 Vehicles, for FMVSS No. 201 Vehicles Without Head-Protection Air Bags, CDS 1995-2009, Chi-Squares Adjusted Downwards Due to Clustering at the PSU Level and Case-Weighting

	ALL OCCUPANTS				BELTED OCCUPANTS			
	Weighted CDS		Unweighted CDS		Weighted CDS		Unweighted CDS	
	Reduction (%)	Rao-Scott χ^2	Reduction (%)	Rao-Scott χ^2	Reduction (%)	Rao-Scott χ^2	Reduction (%)	Rao-Scott χ^2
OVERALL	32	7.67	20	3.37	36	3.91	19	1.67
BELT USE								
Not belted	29	2.18	25	4.19			-48	.48
Belted	36	3.91	19	1.67			15	.78
IMPACT TYPE								
Frontal	30	.75	3	.04	13	.18	-19	
Side impact	20	1.28	16	2.48	25	1.28	15	.78
Rollover first event	58	6.21	42	2.71	62	6.21	57	5.08
Rollover, subsequent worst event	32	.55	33	1.51	40	.31	none	.00
Rear impact or other type	67	1.96	51	.96	-38	.05	42	.30
VEHICLE TYPE								
Passenger car	29	4.37	20	2.49	36	2.55	25	2.20
LTV	37	3.68	20	1.63	42	6.51	13	.25
SEAT POSITION								
Driver	27	3.73	21	2.71	30	1.71	24	1.88
Front passenger	52	10.28	9	.20	59	5.98	-12	.30
Rear passenger	19	.30	17	1.90	11	.02	41	1.14
AGE								
0-12	45	.63	29	.34	-9	.01	-37	.20
13-54	34	10.88	22	3.14	38	3.18	25	2.13
55+	20	.67	4	.05	35	1.99	4	.02
GENDER								
Male	41	7.78	16	1.06	36	2.56	12	.46
Female	15	.79	28	6.19	38	2.43	29	3.14
SPECIFIC HEAD INJURY SOURCE								
A-pillar	41	8.75	20	2.28	45	2.57	34	1.07
B- or other pillars	-7	.07	7	.19	14	.29	14	.42
Header, front or rear	35	1.29	13	.14	11	.02	-49	.47
Side roof rails	41	2.23	19	.54	39	.81	21	.29
Roof interior	59	4.79	37	5.08	55	5.48	32	1.97
SPECIFIC AIS 4-to-6 INJURY TYPE								
Concussion	58	7.38	46	4.18	12	.04	35	1.55
Brain contusion	40	2.57	none	.00	55	6.84	7	.05
Brain, unknown injury	26	3.15	18	2.18	36	3.21	20	1.13
Skull fracture	19	.56	5	.09	26	.78	2	.02

Belted versus unrestrained occupants. The CDS variable MANUSE (or ABELTUSE in the case of automatic 3-point belts) permits separate analyses of belted occupants (including children in safety seats) and unrestrained occupants. The few cases with unknown belt use are excluded from these analyses. Table 2-3 indicates fairly similar effectiveness for both groups. In the weighted data, the 36-percent reduction for the belted occupants is statistically significant, whereas the 29-percent reduction for the unrestrained occupants is not. But the unweighted data are the other way around, with a significant 25-percent reduction for the unrestrained, but only 19 percent for the belted.

By impact type. First-event rollovers are the one impact type with large and consistently significant effects for FMVSS No. 201, for all occupants and belted occupants, with weighted and unweighted data. The injury reduction rises as high as 62 percent for belted occupants in the weighted data. Observed results are generally positive but not statistically significant for the other impact types (frontal, side, subsequent-rollover, rear).

By vehicle type. FMVSS No. 201 may be about equally effective in cars and LTVs. Overall, in the weighted data, the reduction is 37 percent in LTVs and 29 percent in cars, but only the latter is significant at the two-sided .05 level. For belted occupants, effectiveness reaches 42 percent in LTVs, significant at the two-sided .05 level, versus 36 percent in cars. However, in the unweighted data, the effect in cars is greater than or equal to the effect in LTVs.

By occupant seat position. 67 percent of the occupants are drivers; given the limited data on front and rear passengers, effectiveness estimates are less accurate. The effect of FMVSS No. 201 is consistent for drivers, the point estimates of injury reductions ranging only from 21 to 30 percent. Front passengers show higher effectiveness in the weighted data, but a lower or even a negative effect in the unweighted data. All estimates for rear passengers are positive. The results are consistent with a hypothesis that FMVSS No. 201 is about equally effective at the various seat positions.

By occupant age group. Only 5 percent of the occupants are children age 12 or younger, whereas 19 percent are seniors age 55 and older. FMVSS No. 201 is clearly effective for the large intermediate group of occupants age 13-54. Given the limited data, it is unclear if FMVSS No. 201 is also effective for children or seniors; the results are a mix of strong (but non-significant) positives, near-zero effects, and negatives. It is understandable that FMVSS No. 201 might have limited benefits for restrained child passengers, because they are unlikely to reach the upper interior, other than perhaps the B-pillar in a side impact.

By occupant gender. FMVSS No. 201 appears to be effective for both males and females, without being obviously more effective for one than for the other. For all occupants, with weighted data, effectiveness is 41 percent for males (significant at the two-sided .05 level) versus 15 percent for females. But in the other three comparisons, the observed effectiveness is higher for females, and reaches two-sided significance in the unweighted data for all occupants.

By specific injury source. CDS data identifies the injury source and makes it possible to estimate the reduction in head injuries from individual upper-interior sources. These sources are combined into five major groups (infrequent sources such as "windshield plus" are omitted from the analyses):

A-pillar (injury source codes 53 or 103);
B- or other pillars (54, 55, 104 or 105);
Header, front or rear (201 or 202);
Roof side rail (203 or 204); and
Roof interior (205, 206 or 207).

The analysis of A-pillar injuries for all occupants with weighted data, for example, constructs a 2x2 table similar to that in Table 2-2W – except that the left column, "head-upper interior injuries," is limited to the injuries specifically attributed to the A-pillar (injury source codes 53 or 103). But the right column, the control group of "all other AIS 4-to-6 injuries" is identical to Table 2-2W. It still includes all head injuries from sources other than the upper interior and all injuries from any source to the torso, arms or legs – and it still excludes head injuries from upper interior sources other than the A-pillar.

Table 2-3 shows positive effects for FMVSS No. 201 at three locations: the A-pillar (including a statistically significant 41-percent reduction for all occupants in the weighted data); the side roof rails (non-significant point estimates ranging from 19 to 41%); and the roof interior (estimates ranging from 32 to 59 percent, three of them statistically significant). Results for the B-pillar and front/rear header, on the other hand, were not consistently positive. The findings are remarkably consistent with the reductions of average HIC in headform impacts after FMVSS No. 201, which showed large improvements at exactly those locations.[38]

	Average HIC Test Improvement
A-pillar	475.5
B- or other pillar	80.8
Header, front or rear	- 35.2
Roof side rail	245.3
Roof interior	278.8

By specific head-injury type CDS data have several variables to identify a specific type of injury. Based on the body region, lesion and system/organ variables, the four most common head injuries at the AIS 4-to-6 level are:

Concussion (region = head, lesion = concussion, system/organ = brain);
Brain contusion (region = head, lesion = contusion, system/organ = brain);
Brain, unknown injury (region = head, lesion = unknown, system/organ = brain); and
Skull fracture (region = head, lesion = fracture, system/organ = skeletal).

The analysis of concussions for all occupants with weighted data, for example, constructs a 2x2 table similar to that in Table 2-2W – except that the left column, "head-upper interior injuries," is limited to AIS 4-to-6 concussions attributed to upper-interior sources. But the right column, the control group of "all other AIS 4-to-6 injuries" is identical to Table 2-2W. It still includes all head injuries, even concussions, from sources other than the upper interior and all injuries from

---

[38] Kahane & Tarbet (2006), p. 20.

any source to the torso, arms or legs – and it still excludes head injuries from upper interior sources that are not concussions.

Table 2-3 shows an exceptionally strong reduction of concussions after FMVSS No. 201. For all occupants, the observed reduction of concussions in the weighted data is 58 percent, statistically significant at the two-sided .05 level. What makes the result interesting is that FMVSS No. 201 tests HIC in headform impacts, and HIC was originally developed by Versace in the 1960s as a predictor of the risk of concussion, a refinement of the even earlier Wayne State Concussion Tolerance Curve.[39] The latest research continues to demonstrate the correlation of HIC with the risk of concussion.[40] It is reassuring that FMVSS No. 201 shows the highest effectiveness on the type of injury most closely associated with the objective of the standard. However, results were generally positive as well for the other types of AIS 4-to-6 head injury, including a significant 55-percent reduction of brain contusions for belted occupants and two reductions of unknown brain injury that reached significance at the one-sided .05 level.

2.4 Effect on AIS 3 to 6 for specific impact, vehicle, occupant and injury types

Table 2-4 estimates the post-FMVSS No. 201 reductions of AIS 3-to-6 head injuries due to upper-interior contact for various impact, vehicle, occupant, and injury types. Table 2-4 is organized the same way as Table 2-3.[41] The first two estimates reiterate Tables 2-1W and 2-1U, namely, a non-significant overall 11-percent reduction of AIS 3-to-6 head injuries due to upper-interior contact, with a corresponding non-significant 7-percent reduction in the unweighted data.

The principal difference between Tables 2-3 and 2-4 is that FMVSS No. 201 has much higher effectiveness against head injuries at the AIS 4-to-6 level than at the AIS 3-to-6 level. A favorable interpretation is that the technology is primarily effective against the more severe, life-threatening types of head injury (or that it reduces many of the AIS 4-to-6 injuries to AIS 3). That interpretation would be consistent with the next chapter's findings of significant head-injury reduction in fatal crashes. Less favorably, the AIS 3-to-6 results could be a caveat or reality-check on the AIS 4-to-6 results, which perhaps were so positive by chance.

---

[39] Gurdjian, E. S., Webster, J. E., & Lissner, H. R. (1955). Observations on the Mechanism of Brain Concussion, Contusion and Laceration, *Surgery, Gynecology and Obstetrics*, Vol. 101, pp. 680-690; Gadd, C. W. (1966). Use of Weighted-Impulse Criterion for Establishing Injury Hazard, *Proceedings of the Tenth Stapp Car Crash Conference*. New York: Society of Automotive Engineers; Versace, J. (1971). A Review of the Severity Index, *Fifteenth Stapp Car Crash Conference Proceedings*. New York: Society of Automotive Engineers; Newman, J. A., Shewchenko, N., & Welbourne, E. (2000). A Proposed New Biomechanical Head Injury Assessment Function – The Maximum Power Index, *44th Stapp Car Crash Conference*, Paper No. 2000-01-SC16. (Publication No. P-362). Warrendale, PA: Society of Automotive Engineers.

[40] Viano, D. C., Casson, I. R., & Pellman, E. J. (2007). Concussion in Professional Football: Biomechanics of the Struck Player, *Neurosurgery*, Vol. 61, No. 2, pp. 313-328.

[41] The left half of the table estimates effectiveness for all occupants, belted and unbelted. The right half is limited to people who wear seat belts. For each group, Table 2-4 shows the percent injury reduction and the Rao-Scott χ^2 statistic. Reductions that are statistically significant at the two-sided .05 level ($\chi^2 > 3.84$) are printed in bold type and shaded in bright blue; significance at the one-sided .05 level ($\chi^2 > 2.71$) is shaded in pale blue; a blue border indicates χ^2 in the 80-90th percentiles (1.64-2.70). Negative estimates are printed in red (with a red border if χ^2 is in the 80-90th percentiles).

Table 2-4: Reduction (%) of **AIS 3-to-6** Head Injuries Due to Upper Interior Contact, Relative to Pre-FMVSS No. 201 Vehicles, for FMVSS No. 201 Vehicles without Head-Protection Air Bags, CDS 1995-2009, Chi-Squares Adjusted Downwards Due to Clustering at the PSU Level and Case-Weighting

	ALL OCCUPANTS				BELTED OCCUPANTS			
	Weighted CDS		Unweighted CDS		Weighted CDS		Unweighted CDS	
	Reduction (%)	Rao-Scott χ^2	Reduction (%)	Rao-Scott χ^2	Reduction (%)	Rao-Scott χ^2	Reduction (%)	Rao-Scott χ^2
OVERALL	11	.89	7	.45	18	1.18	7	.24
BELT USE								
Not belted	none	.00	11	.75				
Belted	18	1.09	7	.24				
INJURY SEVERITY								
AIS 4-6 (life-threatening)	32	7.67	20	3.37	36	3.91	19	1.67
AIS 3 (not life-threatening)	-5	.12	-5	.14	8	.12	-5	.14
IMPACT TYPE								
Frontal	-12	.20	-11	.47	-12	.18	-26	1.80
Side impact	8	.25	4	.11	15	.23	-3	.04
Rollover first event	13	.86	18	.75	12	.14	40	3.59
Rollover, subsequent worst event	11	.06	10	.13	-30	.13	-43	.51
Rear impact or other type	67	2.90	19	.11	-36	.08	11	.02
VEHICLE TYPE								
Passenger car	none	.00	5	.21	19	.76	8	.28
LTV	26	2.46	11	.59	18	1.07	10	.20
SEAT POSITION								
Driver	3	.05	10	.67	13	.46	12	.55
Front passenger	30	2.96	-6	.12	38	1.96	-14	.63
Rear passenger	15	.29	13	.38	-3	.00	14	.14
AGE								
0-12	39	.43	19	.13	6	.01	-26	.12
13-54	6	.33	8	.53	7	.15	12	.83
55+	18	.71	-3	.04	40	2.94	-7	.06
GENDER								
Male	26	3.85	5	.13	9	.20	none	.00
Female	-18	.40	10	.65	31	1.27	16	1.08
SPECIFIC HEAD INJURY SOURCE								
A-pillar	29	1.07	13	.96	54	2.26	38	4.17
B- or other pillars	5	.07	6	.18	17	.54	6	.10
Header, front or rear	2	.00	-9	.06	-61	.58	-89	1.74
Side roof rails	20	.66	1	.00	2	.02	-3	.04
Roof interior	27	3.63	20	1.65	18	.91	17	.68
SPECIFIC AIS 3-to-6 INJURY TYPE								
Concussion	44	4.48	44	6.54	14	.09	41	6.41
Brain contusion	26	2.44	-5	.11	26	1.31	-16	.51
Brain, unknown injury	5	.10	4	.14	15	.83	6	.13
Skull fracture	-11	.51	4	.12	2	.01	1	.01
Facial fracture	2	.00	-10	.16	27	.13	9	.06

Belted versus unrestrained occupants. With weighted data, FMVSS No. 201 has no observed effect for unrestrained occupants and reduces AIS 3-to-6 head injuries by 18 percent for belted occupants. But with the unweighted data, the reductions are 11 percent for unrestrained and 7 percent for belted. None of the effects is statistically significant.

Life-threatening versus AIS 3 injuries. FMVSS No. 201 significantly reduces life-threatening (AIS 4-to-6) head injuries. However, for head injuries with AIS exactly 3 (serious but not life-threatening), Table 2-4 shows no benefit, in fact a non-significant 5 percent increase for all occupants (in the weighted and also in the unweighted CDS data). For belted occupants, there is a non-significant 8-percent reduction with weighted data and a non-significant 5 percent increase with unweighted data.

AIS 2 injuries. The effect of FMVSS No. 201 on head injuries of the next lower level of severity, AIS 2 (moderate, not life-threatening) can be estimated by extending the database created in Section 2.1 to include these injuries. The results are similar to the results for AIS exactly 3, perhaps slightly more favorable. For all occupants, there is a non-significant 5 percent increase in AIS 2 injuries for all in the weighted CDS data, but a non-significant 11-percent reduction in the unweighted data. For belted occupants, there is a non-significant 7-percent reduction with weighted data and a statistically significant 15-percent reduction with unweighted data.

By impact type. Paralleling Table 2-3, results are more favorable for first-event rollovers than other types of crashes, but none of the AIS 3-to-6 analyses show unequivocal benefits for FMVSS No. 201. In first-event rollovers, all estimates are positive, but only the unweighted estimate for belted occupants is significant, and only at the one-sided .05 level. All estimates for frontal crashes are negative, but none are significant. The other types of crashes have a mix of positive and negative results

By vehicle type. At the AIS 3-to-6 level, results lean slightly in the direction of being more favorable for LTVs than cars. No estimate is statistically significant.

By occupant seat position. As in Table 2-3, the data do not entail a conclusion that FMVSS No. 201 is more effective at one seat position than at another.

By occupant age group. There is little evidence that FMVSS No. 201 is more effective, or less effective, for children or seniors than for the intermediate group of occupants age 13-54.

By occupant gender. For all occupants, with weighted data, there is a significant 26-percent injury reduction for males and a non-significant 18-percent increase for females. But in the other three analyses, the observed effectiveness is higher for females than for males. The inconstant results do not lead to any conclusion.

By specific injury source. FMVSS No. 201 is consistently beneficial in contacts with the A-pillar (including a significant reduction in the unweighted data for belted occupants) and with the roof interior (including one-sided significance in the weighted data for all occupants). For the other injury sources, estimates are small or inconsistent. The results parallel Table 2-3.

By specific head-injury type Even including AIS 3 injuries, FMVSS No. 201 is quite effective against concussions, with three of the effectiveness estimates exceeding 40 percent and significant at the two-sided .05 level. FMVSS No. 201 does not show a consistent, large benefit for any of the other injury types, including facial fractures (an injury whose maximum AIS is 3).

CHAPTER 3

EFFECT OF FMVSS No. 201 ON HEAD INJURIES IN FATAL CRASHES: ANALYSES OF 1999-2007 FARS-MCOD DATA

3.0 Summary

The FARS-MCOD (multiple cause of death) file enables statistical analyses of the ratio of head injuries to other injuries before and after the 1999-2003 upgrade of FMVSS No. 201. The data does not identify the source of the injuries and do not allow singling out injuries due to upper-interior contact. Nevertheless, the analyses show statistically significant reductions of head injuries, relative to other injuries, after FMVSS No. 201: 6 percent for all fatally-injured occupants and 10 percent for belted occupants. They parallel the similarly positive findings of the preceding chapter's analyses of CDS data. FMVSS No. 201 is especially effective in first-event rollovers, also as in the preceding chapter; however, unlike that chapter, FMVSS No. 201 is especially effective in LTVs, for belted occupants, and for female occupants here.

3.1 Injuries contributing to occupant fatalities, before and after FMVSS No. 201

FARS is a census of fatal crashes in the United States since 1975. The basic FARS data does not furnish information on the specific injuries of people involved in the crashes. The National Center for Health Statistics, however, has assembled a census of death certificates for people who died in the United States since 1968, from any causes.[42] Its data is called the Multiple Cause of Death file.[43] Death certificates list diseases, injuries, conditions and external factors that "contributed" to a person's death. Beginning with the 1987 data, NHTSA and NCHS have worked together to link records of fatalities on FARS to their corresponding death-certificate data on MCOD. These supplemental FARS-MCOD files can be merged with the basic FARS person-level data by ST_CASE, VEH_NO and PER_NO, but only for the fatally injured people. They list the injuries of the fatally injured people – not necessarily all their injuries, but only those that "contributed to the fatality" in the opinion of whoever filled out the death certificate.

Injury data. MCOD uses the International Classification of Diseases to classify injuries. Since 1999, MCOD has used the 10th revision of that system (ICD-10).[44] It differs substantially from earlier versions. That makes 1999 a good starting point for the data in our analyses, because it is also the first model year that any vehicles certified to FMVSS No. 201. Because acquisition and processing of death-certificate data takes time, 2007 is the latest year of FARS-MCOD data as of May 2011. The analysis is based on nine years of data, 1999 to 2007.

An ICD-10 code consists of a letter, a two-digit number and, possibly, a decimal point and another digit. For example, S02.0 is a fracture of the vault of the skull. Relevant to this study

---

[42] CDC. *Multiple Cause of Death, 1999-2006*. http://wonder.cdc.gov/wonder/help/mcd.html.
[43] NCHS (2006). *Multiple Causes of Mortality, 2003; Documentation of the Mortality Tape File for 2003 Data*. Hyattsville, MD: National Center for Health Statistics.
http://wonder.cdc.gov/wonder/sci_data/mort/mcmort/type_txt/mcmort03/mcmort03.asp.
[44] WHO (2005). *International Statistical Classification of Diseases and Health Related Problems, Tenth Revision – ICD-10, Second Edition*. Geneva: World Health Organization. http://www.who.int/classifications/icd/en/.

are the codes starting with S and T, comprising all types of injuries as well as poisonings and consequences of trauma. Not relevant are the codes starting with other letters, such as the A-R codes for diseases or the V codes for external causes including "motor vehicle crash."

The FARS-MCOD file may list up to 15 "record-axis" ICD-10 codes per fatally injured person, mostly S and T codes. The record-axis codes recapitulate whatever is on the death certificate, except that NCHS has screened them to eliminate contradictions and duplicate codes. The 1999-2007 FARS-MCOD files have 445,308 S and T codes for 273,287 fatally injured people, an average of 1.6 injuries per person. The majority of people, 170,825 have only one injury listed; 63,576 have two; 23,110 have three; and 15,776, four or more. In other words, most of the death certificates appear to be coded appropriately, listing only the one, two, or occasionally three or four injuries that clearly contributed to the fatality. Far less often is there an extensive list of injuries that likely does not discriminate their importance.

The basic analysis approach is to count the number of head injuries (which can potentially be reduced by FMVSS No. 201) and compute its ratio to a control group of injuries that are unlikely to be affected by FMVSS No. 201. The ratios are compared for pre- and post-FMVSS No. 201 vehicles.

However, injury information on death certificates does not necessarily derive from autopsies or hospital records, and it may lack specifics needed even for basic analysis. The priority for many of the officials who fill out original death certificates is to register whether a person died from an unintentional crash as opposed to homicide, suicide or a heart attack – not what particular injury caused the fatality. The most common code, accounting for 20 percent of the reported injuries (90,981 of 445,308) is simply T07, "unspecified multiple injuries." Also prevalent are T14.9, "injury, unspecified" (42,845 cases) and T14.8, "other injuries of unspecified body region" (7,371 cases). These cases are lost to our analysis.

Even when the codes specify a body region and can be used for our basic analyses, they often say little else. For example, code S09.9, "unspecified injury of head" (86,218 cases) exceeds all other head injuries, combined. The small reported numbers of specific injuries will make it impossible to conduct more detailed analyses of the effect of FMVSS No. 201 on specific injury types.

Appendix B lists all the S and T codes, specifying which ones are counted as head injuries in our analysis (red print), as control-group injuries (blue print) or not at all (black print). The control group of injuries unlikely to be affected by FMVSS No. 201 includes injuries to the torso, arms or legs (including a combination of torso, arms and/or legs), plus all burns, drowning, asphyxiation and poisoning by carbon monoxide and other substances intrinsic to a motor vehicle.

Neck injuries are excluded from the analyses, because it is unclear if FMVSS No. 201 might have an effect. Also excluded are the many injuries with unknown body region or unspecified multiple regions, plus poisoning by drugs or other substances not intrinsic to a motor vehicle, and delayed effects or complications. Furthermore, the analyses exclude the large number of codes (but not too many cases) of relatively minor injuries that would not ordinarily "contribute to death" but were nevertheless listed on death certificates – e.g., S92.5, "fracture of non-big toe"; more generally, single fractures of the arm or lower leg or superficial wounds. These codes are listed as "minor" in Appendix B. However, unspecified injuries to a specific body region cannot be excluded, even though at least some of them are undoubtedly minor, because they account for such a large share of the data.

The MCOD data does not list the sources (contact points) of the injuries. We cannot differentiate head injuries due to contacts with the upper interior from other head injuries. Therefore, the analysis approach is limited to counting the number of head injuries (which include head injuries from upper interior sources as well as head injuries from other sources) and computing its ratio to the control group: injuries that are not head injuries (and these injuries hardly ever involve upper interior contact). The ratios are compared for the fatally injured occupants of pre- and post-FMVSS No. 201 vehicles, including passenger cars and LTVs (pickup trucks, SUVs and vans with less than 10,000 pounds GVWR).

FARS-MCOD analysis file Section 2.1 developed a file for studying the effect of FMVSS No. 201 in the CDS. Because that data had a fine analytic tool – it identified head injuries specifically due to upper interior contact – we were at liberty to include many model years (1985-2009). The only restrictions were that the data had to be uniform on a few factors that greatly affected the risk of control group injuries, namely, the presence of frontal air bags in frontal crashes and the absence of side air bags in non-frontal crashes.

But FARS-MCOD data only identifies head injuries. It does not identify whether these head injuries are due to upper interior contact. The potential effect of FMVSS No. 201 on **all** head injuries is a fraction (43% according to Section 1.1) of its potential effect on head injuries specifically due to upper interior contact. But as the signal is smaller, the noise increases. More factors could affect overall head injury than just the factors that affect head injuries due to upper interior contact.

When selecting vehicles for the analysis, we must strive to make the group of FMVSS No. 201-certified vehicles as similar as possible to the group of pre-standard vehicles (except for the change in FMVSS No. 201 certification). Here are the restrictions imposed to make the FARS-MCOD database as uniform as possible. Restrictions that already applied to the CDS database in Section 2.1 are printed in Italic type. New, more stringent restrictions are printed in regular block letters:

- *As described in Section 2.1,* Buying a Safer Car *and its supporting manufacturer letters identify if makes and models initially certified to FMVSS No. 201 in model year 1999, 2000, 2001, 2002, or 2003.*[45]

- This analysis is limited to makes and models produced both before and after FMVSS No. 201 certification.

---

[45] NHTSA. (1998). *New Car Safety Features 1999.* (Report No. DOT HS 808 808). Washington, DC: National Highway Traffic Safety Administration; NHTSA. (2000). *Buying a Safer Car 2000.* (Report No. DOT HS 809 046). Washington, DC: National Highway Traffic Safety Administration; NHTSA. (2001). *Buying a Safer Car 2001.* (Report No. DOT HS 809 152). Washington, DC: National Highway Traffic Safety Administration; NHTSA. (2002). *Buying a Safer Car 2002.* (Report No. DOT HS 809 409). Washington, DC: National Highway Traffic Safety Administration.

- Skip any model year where FMVSS No. 201 certification is unknown or doubtful. For example, if a vehicle first certified in 1999 (the first phase-in year), but was substantially redesigned in 1998, do not use model year 1998 in the analysis, as it is possible that the 1998 vehicles already had the same design features as the 1999s.

- *All vehicles equipped with head-protection air bags at any seat position are excluded from the analysis, regardless of whether these air bags are the curtain, inflatable-tube or head-torso combination type and regardless of whether the air bags deployed.*

- Specifically, the analysis is limited to makes and models for which at least some of the FMVSS No. 201 certified vehicles did not have head-protection air bags.

- At a maximum, include the last three model years before FMVSS No. 201 certification and the first three FMVSS No. 201-certified years. For example, if a model initially certified in 2002, include up to 1999-2001 (pre-standard) and 2002-2004 (post-standard).

- Of course, if a make and model began production less than three years before FMVSS No. 201 certification, or ended less than three years afterwards, we are immediately limited to the years it was produced.

- "Starting or ending production" would also include replacing a model with a radically different vehicle of the same name (e.g., Volkswagen Beetle and New Beetle), but not the customary restyling where a vehicle remains in the same functional class with fairly small changes in wheelbase or appearance.

The maximum span of six model years for a specific make and model is likewise abbreviated if there are any of the following major changes in that model's safety systems during that time:

- Exclude any passenger car, SUV, van or crew-cab pickup truck not equipped with dual air bags. Exclude any pickup truck without a crew cab unless it was equipped with dual air bags plus an on-off switch for the passenger.

- Exclude any passenger car not certified to FMVSS No. 214. (FMVSS No. 214 phased into passenger cars in model years 1994-1997.) The exclusion is unnecessary for LTVs, because most LTVs needed little or no change to meet FMVSS No. 214.[46]

- Vehicles of the same make and model must be consistent on availability of side air bags for torso protection: all of them are equipped, or none. If torso bags are VIN-identifiable options in some model years, just exclude the vehicles equipped with them. If FMVSS No. 201 certification coincided with standard torso bags, exclude that model from the analysis.

- In the analyses that include **belted** occupants, vehicles of the same make and model must be consistent on availability of seat-belt pretensioners: all of them are equipped, or none. If FMVSS No. 201 certification coincided with the installation of pretensioners, eliminate that model from analyses that include belted occupants. However, for analyses limited to

---

[46] Kahane (2007), pp. 33-34.

unbelted occupants, pretensioners are irrelevant and are not a basis for excluding a model year or an entire make and model.[47]

Finally, model years are deleted from any make and model if that is necessary to **balance** the sample between original and sled-certified air bags – i.e., to assure more or less the same make-model mix in the pre-standard and the post-standard vehicles. Thus, for example, if for any of the preceding reasons, we can only use the first model year after FMVSS No. 201 for a particular make and model, we will also use only the last model year before FMVSS No. 201. (However, because the earlier vehicles are on the road longer and have accumulated more data, we can allow one extra post-standard model year if we are limited to just one or two pre-standard model years.[48])

Occupants may be of any age and in any seat position.

Appendix C of this report lists every make and model that is a candidate for possible inclusion in the analyses and specifies the model year when FMVSS No. 201 certification began. It summarizes the availability of other safety equipment (side air bags, frontal air bags, pretensioners, FMVSS No. 214 certification) before and after FMVSS No. 201. Based on those criteria, it specifies the range of model years, if any, used in analyses that include belted occupants and in analyses limited to unbelted occupants – and explains why.

The analysis file includes 24,570 individual injuries of known body region that "contributed to the death of an occupant," listed in red print (head injuries) or blue print (control group injuries) in Appendix B (and excluding neck injuries, minor injuries and other categories listed in black print in Appendix B). The file includes 14,875 injuries in pre-standard vehicles and 9,695 in FMVSS No. 201-certified vehicles. The 24,570 individual injuries were reported for 15,130 fatally injured occupants of 13,986 distinct vehicles involved in 13,958 separate crashes.

3.2 Overall reduction of head injuries

Contingency table analysis. Table 3-1 shows 8,002 head injuries and 6,873 injuries to other body regions (torso, arms, legs or a combination of those) before FMVSS No. 201, a risk ratio of 1.164. In the FMVSS No. 201-certified vehicles without head-protection air bags, there were 5,055 head injuries and 4,640 other injuries, a risk ratio of 1.089. That is a 6-percent reduction of head injuries relative to the control group.

---

[47] However, we do not abbreviate the model years upon introduction of load limiters for safety belts because (1) the effect is probably smaller; (2) this would exclude a large proportion of the data; and (3) it is not certain when load limiters were introduced in many of the makes and models. See also Walz, M. C. (2003). *NCAP Test Improvements with Pretensioners and Load Limiters*. (Report No. DOT HS 809 562). Washington, DC: National Highway Traffic Safety Administration. Available at http://www-nrd.nhtsa.dot.gov/Pubs/809562.PDF.

[48] See also Kahane, C. J. (1996). *Fatality Reduction by Air Bags*. (Report No. DOT HS 808 470). Washington, DC: National Highway Traffic Safety Administration, pp. 7-9. Available at http://www-nrd.nhtsa.dot.gov/Pubs/808470.PDF.

TABLE 3-1

OVERALL EFFECT OF FMVSS No. 201 ON HEAD INJURIES
THAT "CONTRIBUTED TO DEATH," PASSENGER CARS AND LTVs
(1999-2007 FARS-MCOD)

	Head Injuries	Other Injuries	Risk Ratio	Reduction (%)
Pre-FMVSS No. 201	8,002	6,873	1.164	
FMVSS No. 201-certified	5,055	4,640	1.089	6

The SURVEYFREQ procedure of SAS analyzes contingency tables such as Table 3-1 and generates a Rao-Scott chi-square (χ^2) statistic. It has the same critical values as a conventional χ^2, namely 3.84 is its 95th percentile (the minimum amount needed for statistical significance at the two-sided .05 level), 2.71 is its 90th percentile (significance at the one-sided .05 level), and 1.64 is its 80th percentile. The procedure first calculates the conventional χ^2 if the cell counts in Table 3-1 had been independent observations from a simple random sample. However, these counts of injuries cannot be considered fully independent observations because:

- The same occupant may have more than one injury that "contributed to death" according to FARS-MCOD; thus, the injury cases are not necessarily independent observations. As noted above, the 24,570 injuries were distributed among 15,130 occupants.

- Likewise, one vehicle may have more than one fatally injured occupant, and one crash may involve more than one vehicle from our selected make-model groups and model-year ranges. The 24,570 injuries were distributed among 13,986 vehicles involved in 13,958 separate crashes.

The 24,570 injuries in Table 3-1 may be construed as a "cluster" sample, with each of the 13,958 separate crashes constituting a "primary sampling unit." The Rao-Scott χ^2 is the SRS χ^2 is divided by the design effect (DEFF) due to the numbers deriving from a cluster sample of 13,958 PSUs. The Rao-Scott χ^2 statistic for Table 3-1 is 4.56. In other words, the 6--percent reduction of head injuries after FMVSS No. 201 is statistically significant at the two-sided .05 level.

Logistic regression analysis allows an adjustment for some imbalances and trends in the basic data. Above all, as occupants age, they become increasingly vulnerable to chest injuries and, as a consequence, their ratio of head injuries to other injuries decreases. If vehicles certified to FMVSS No. 201, being newer vehicles, had a slightly smaller proportion of older occupants, it would increase their ratio of head injuries to other injuries and possibly mask the benefit of FMVSS No. 201. Logistic regression also adjusts for an over- or under-representation of certain makes and models in the pre- or post-FMVSS No. 201 groups.

The 24,041 data points in the regression are the FARS-MCOD injury cases from Table 3-1 (but excluding 529 of the original 24,570 cases where the impact type or the occupant's age or gender were not reported). The dependent variable, HEAD_INJ equals 1 for head injuries, 2 for injuries to other body regions. The key independent variable, FMVSS201 equals 0 for pre-standard vehicles, 1 for post-standard. The other independent variables are:

- The occupant's age, entered directly as a linear variable.
- The occupant's gender, entered as a categorical variable.
- The occupant's seat position, entered as a categorical variable (driver, front-seat passenger, back-seat passenger).
- The vehicle's make and model, entered as a categorical variable (65 separate makes and models, generally corresponding to the models listed in Appendix C; however, a few sparse-data models have been combined with others[49]).
- Vehicle age, entered directly as a linear variable.
- The impact type, entered as a categorical variable (first event rollover, frontal, side impact, rear/other).

The SURVEYLOGISTIC procedure of SAS performs logistic regressions with a mix of linear and categorical independent variables and generates a Wald chi-square (χ^2) statistic that has been adjusted for the design effect of cluster sampling, namely, that the injury cases are clustered at the crash level (and each crash is a "PSU"). The coefficient for FMVSS201 is -.0611. In other words, FMVSS No. 201 certification (without head-protection air bags) is associated with a 1 - exp(-.0611) = 6-percent reduction in head injuries relative to other injuries – identical to the result of the simple contingency-table analysis of Table 3-1. The coefficient is statistically significant at the one-sided .05 level, as evidenced by a χ^2 = 3.18 – and this χ^2 statistic is adjusted for the design effect. (The coefficient for AGE was -.0156, with a χ^2 of 361.8: the ratio of head injuries to other injuries decreases by 1.56 percent for each year that a person ages.)

3.3 Belted versus unrestrained occupants

Contingency table analyses. Based on the FARS variable REST_USE, we can separately analyze belted occupants (including children in safety seats) and unrestrained occupants. Cases with unknown belt use are excluded from these analyses. Furthermore, as explained in Section 3.1 and Appendix C, we can include additional makes and models/model years in the analyses of unrestrained occupants, because a change in the availability of seat belt pretensioners is not an issue. Table 3-2 shows a statistically significant 10-percent reduction in head injuries after FMVSS No. 201 for the belted occupants, versus a non-significant 4-percent reduction for the unrestrained occupants.

---

[49] Dodge Viper, Mazda Miata, Mercedes SLK, and Acura NSX with Chevrolet Corvette; Acura RL, with Acura TL; and Toyota Prius with Toyota Corolla.

TABLE 3-2

EFFECT OF FMVSS No. 201 ON HEAD INJURIES
THAT "CONTRIBUTED TO DEATH," PASSENGER CARS AND LTVs
BY OCCUPANTS' BELT USE
(1999-2007 FARS-MCOD)

	Head Injuries	Other Injuries	Risk Ratio	Reduction (%)
BELTED OCCUPANTS (including child passengers in safety seats)				
Pre-FMVSS No. 201	3,172	2,860	1.109	
FMVSS No. 201-certified	1,990	1,993	.998	10
UNRESTRAINED OCCUPANTS				
Pre-FMVSS No. 201	6,729	5,312	1.267	
FMVSS No. 201-certified	3,882	3,188	1.218	4

Rao-Scott χ^2 is 4.82 for the table of belted occupants, indicating that the injury reduction is statistically significant at the two-sided .05 level. The χ^2 is a non-significant 1.25 for the table of unrestrained occupants.

Logistic regression analyses again generate similar results. For belted occupants, the coefficient for FMVSS201 is -.1058. In other words, FMVSS No. 201 certification (without head-protection air bags) is associated with a $1 - \exp(-.1058) =$ 10-percent reduction in head injuries relative to other injuries, identical to Table 3-2. The coefficient is statistically significant at the two-sided .05 level, as evidenced by a Wald χ^2, adjusted for the design effect of 3.92. The coefficient for FMVSS201 for unrestrained occupants is .0017. In other words, FMVSS No. 201 certification is associated with a non-significant ($\chi^2 = .002$), near-zero effect on head injuries for unrestrained occupants.

The 10-percent reduction for belted occupants is consistent with the CDS analyses of Section 2.3. Because approximately 43 percent of life-threatening head injuries have upper-interior sources (as discussed in Section 1.1), the 10-percent reduction of head injury in FARS-MCOD is equivalent to a 23-percent reduction of head injury due to upper-interior contact. The corresponding reductions of life-threatening (AIS 4-to-6) head injuries due to upper-interior contact for belted occupants in CDS (Table 2-3) were 36 percent in the weighted data and 19 percent in the unweighted data: 23 percent is in between them. On the other hand, CDS shows about the same effectiveness for unrestrained as for belted occupants, whereas FARS-MCOD has little evidence of effectiveness for unrestrained occupants.

3.4 Effect for specific impact, vehicle and occupant types

Table 3-3 estimates the post-FMVSS No. 201 reductions of head injuries that "contributed to death" for various impact, vehicle, occupant, and injury types. The left half of the table estimates effectiveness for all occupants, belted and unbelted. The right half is limited to people who wear seat belts.

TABLE 3-3

REDUCTION (%) OF HEAD INJURIES THAT "CONTRIBUTED TO DEATH," RELATIVE TO PRE-FMVSS No. 201 VEHICLES, FOR FMVSS No. 201 CARS AND LTVs WITHOUT HEAD-PROTECTION AIR BAGS, FARS-MCOD 1999-2007
Chi-Squares Adjusted Downwards Due to Clustering at the Case Level

	ALL OCCUPANTS		BELTED OCCUPANTS	
	Reduction (%)	Rao-Scott χ^2	Reduction (%)	Rao-Scott χ^2
OVERALL	6	4.56	10	4.82
BELT USE				
Not belted	4	1.25		
Belted	10	4.82		
IMPACT TYPE				
Frontal	1	.02	8	1.20
Side impact	1	.06	2	.10
Rollover first event	21	8.70	30	5.60
Rear impact or other type	14	1.76	21	1.62
VEHICLE TYPE				
Passenger car	3	.41	6	.95
LTV	11	6.63	18	6.70
SEAT POSITION				
Driver	8	5.33	10	3.26
Front passenger	-6	.58	8	.65
Rear passenger	15	3.09	23	2.04
AGE				
0-12	23	2.43	36	4.06
13-54	6	3.30	12	4.49
55+	3	.21	2	.07
GENDER				
Male	1	.03	8	1.62
Female	16	11.61	13	3.90

For each group, Table 3-3 shows the percent injury reduction and the Rao-Scott χ^2 statistic. Reductions that are statistically significant at the two-sided .05 level ($\chi^2 > 3.84$) are printed in bold type and shaded in bright blue; significance at the one-sided .05 level ($\chi^2 > 2.71$) is shaded in pale blue; a blue border indicates χ^2 in the 80th to 90th percentiles (1.64-2.70). Negative estimates are printed in red (none are significant). The first three rows reiterate Tables 3-1 and 3-2, namely, a statistically significant overall 6-percent reduction head injuries, reaching 10 percent for belted occupants but only 4 percent for unrestrained occupants.

Table 3-3 tries to define the same categories as Table 2-3, the corresponding analysis of AIS 4-to-6 injuries in CDS: by belt use, impact type, vehicle type, seat position, occupant age, and gender. But it does not include the last two sets of analyses: by injury source and by specific injury type. FARS-MCOD does not report injury sources within the vehicle. Even though the ICD-10 codes potentially include detailed injury descriptions, the information on death certificates does not necessarily derive from autopsies or hospital records and more often than not lacks detail, as explained in Section 3.1. For example, code S09.9, "unspecified injury of head" exceeds all other head injuries, combined. The small reported numbers of specific injuries make it impossible to analyze the effect of FMVSS No. 201 on specific injury types.

By impact type. First-event rollovers are the one impact type with large, statistically significant benefits for FMVSS No. 201, for all occupants and belted occupants, reaching 30 percent for the belted occupants. Observed results are positive but not statistically significant for the other impact types (frontal, side, rear/other). The results are entirely consistent with CDS (Table 2-3).

By vehicle type. Table 3-3 shows strong, statistically significant head-injury reductions in LTVs, 11 percent for all occupants and 18 percent for belted occupants, but much smaller, non-significant reductions in passenger cars, 3 percent and 6 percent, respectively. The trend is directionally consistent with the reductions of average HIC in headform impacts after FMVSS No. 201. In LTVs HIC averaged 956 in the impacts to pre-standard interiors and fell to 655 after FMVSS No. 201 certification, an average improvement of 301 units. In passenger cars, pre-standard HIC was not so high, averaging 855, and fell to 683, a somewhat smaller (but still significant) average improvement of 172 units.[50] But it differs from the CDS results, which showed approximately equal effectiveness for FMVSS No. 201 in cars and LTVs.

By occupant seat position. Sixty-nine percent of the occupants are drivers; given the limited data on front and rear passengers, effectiveness estimates are less accurate. The effect of FMVSS No. 201 is consistently positive for drivers. Front passengers show a non-significant negative effect when all are included (belted and unrestrained), but a positive effect, similar to drivers, for just the belted passengers. The estimates for rear passengers are quite positive. The results are consistent with CDS and provide little basis to reject a hypothesis that FMVSS No. 201 is about equally effective at the various seat positions.

By occupant age group. Only 4 percent of the occupants are children 12 or younger, whereas 22 percent are seniors 55 and older. FMVSS No. 201 is effective for the large intermediate group of occupants 13 to 54. In the limited data, FMVSS No. 201 looks effective for age 0-12, reaching a statistically significant 36-percent head-injury reduction for restrained children (hard to explain, since restrained children would not easily contact the upper interior). There is little

---

[50] Kahane & Tarbet (2006), p. 23.

observed benefit for seniors. The results are somewhat ambiguous about relative FMVSS No. 201 effectiveness, but not inconsistent with the CDS findings.

By occupant gender. Table 3-3 shows a strong, statistically significant 16-percent head-injury reduction for all female occupants, belted plus unrestrained, versus only 1 percent for male occupants. For just belted occupants, the contrast is not quite as great, but Table 3-3 still shows a significant 13-percent reduction for females and a non-significant 8-percent reduction for males. There does not appear to be any obvious intuitive explanation why FMVSS No. 201 should be so effective for females, especially unrestrained females. Furthermore, the CDS analyses did not show any strong differences in effectiveness for males and females.

Reconciliation with CDS. The FARS-MCOD analyses and the CDS results for AIS 4-to-6 injuries both show a statistically significant overall reduction of head injuries for FMVSS No. 201 (without curtain bags); moreover, the results are quantitatively consistent, taking into account that FARS-MCOD measures the reduction in all types of head injury while CDS identifies injuries specifically due to upper interior contact. The databases also agree that effectiveness is especially high in first-event rollovers. The CDS findings that FMVSS No. 201 is especially effective against concussions and in contacts with the A-pillar or roof interior are consistent with the definition of HIC and the HIC improvements in impact tests of FMVSS No. 201-certified versus pre-standard vehicles. All of the preceding supports a fairly strong conclusion that FMVSS No. 201 (even without curtain bags) has been effective in reducing life-threatening head injuries.

Nevertheless, a few inconsistencies between the databases – namely that FMVSS No. 201 is especially effective in FARS-MCOD, but not CDS, for LTVs, for belted occupants, and for female occupants – and the much less positive CDS results at the AIS 3-to-6 level than at the AIS 4-to-6 level may be considered caution flags on the results. The effectiveness of FMVSS No. 201 (without curtain air bags) is not completely beyond a reasonable doubt. But with curtain air bags becoming standard in all new vehicles, there will not be much new data arriving to potentially refine the current results.

CHAPTER 4

DISCUSSION: BEST EFFECTIVENESS ESTIMATE, BENEFITS, AND COSTS

4.0 Summary

FMVSS No. 201 without head-protection air bags reduces AIS 4-to-6 head injuries due to contact with upper-interior components by an estimated 24 percent (95% confidence bounds, 11 to 35%), based on the average of the results of the CDS and FARS analyses. That is equivalent to a 4.3-percent reduction of overall fatality risk (confidence bounds 2.0 to 6.2%). When all vehicles on the road meet FMVSS No. 201, it will be saving an estimated 1,087 to 1,329 lives per year. At a cost of $25.52 (in 2010 dollars) over the life of a vehicle, that amounts to an annual cost ranging $301 to $424 million for certifying all new vehicles to FMVSS No. 201. It is a very cost-effective regulation, costing less than $1 million per life saved.

4.1 Best effectiveness estimate and its confidence bounds

The principal effectiveness estimate from the analyses of CDS data (Table 2-2W) is that AIS 4-to-6 head injuries due to contact with upper-interior components decreased by 32 percent after FMVSS No. 201. It is based on weighted CDS data. Although that estimate is statistically significant (Rao-Scott $\chi^2 = 7.67$) and presumably unbiased, there is evidence that it is likely higher than the true effect of FMVSS No. 201, perhaps by chance, because of sampling error associated with the use of weighted data. The first evidence is that the corresponding estimate with unweighted CDS data (Table 2-2U) is only 20 percent. While this lower estimate may be more statistically precise and may give a better idea of the true effect, it is not a useable number. Estimates from CDS are normally based on weighted data because biases may be introduced by failing to weight the CDS data. Additional evidence is the principal estimate from FARS data (Table 3-1), namely that head injuries of fatally injured occupants decreased by a statistically significant 6.43 percent. Taking into account that 43 percent of AIS 4-to-6 head injuries, prior to FMVSS No. 201, were due to upper-interior contact (Table 1-3), that corresponds to a 6.43/.43 = 15-percent reduction of head injuries due to upper-interior contact: well below the weighted-CDS estimate and even somewhat below the unweighted CDS. The average of the weighted-CDS and the FARS results will serve as this report's best estimate of overall effectiveness. It is the average of two statistically significant, unbiased estimates comprising all the data used in this report.

The "average" of the two estimates will be the harmonic average – i.e., it is based on the simple arithmetic average of the log-odds ratios for the two analyses. In Table 2-2W, the CDS analysis, the 32-percent effectiveness estimate corresponds to a log-odds ratio of:

$$\log [(18{,}791/93{,}892) / (103{,}829/354{,}393)] = -.3811$$

Based on Tables 1-3 and 3-1, the 15 percent effectiveness from FARS corresponds to a log-odds ratio of:

$$\log \{1 - [(239{,}534/103{,}865) \times [1 - (5{,}055/4{,}640) / (8{,}002/6{,}873)]]\} = -.1605$$

The average of -.3811 and -.1605 is -.2708, corresponding to an effectiveness estimate of:

$$1 - \exp(-.2708) = 23.73 \text{ percent}$$

Confidence bounds. The CDS effectiveness estimate has Rao-Scott $\chi^2 = 7.67$. The standard deviation of the log-odds ratio is $.3811/\sqrt{7.67} = .1376$. The FARS-based effectiveness estimate is also statistically significant and has Rao-Scott $\chi^2 = 4.56$. The standard deviation of the log-odds ratio is $.1605/\sqrt{4.56} = .0752$. The average of the two effectiveness estimates is a linear combination of two independent statistics. The standard deviation of the average log-odds will be $\sqrt{(.1376^2 + .0752^2)} = .0784$. The 95-percent confidence bounds for the average log-odds are:

$$-.2708 \pm 1.96 \times .0784 = -.4251 \text{ to } -.1171$$

The 95-percent confidence bounds for the best effectiveness estimate are 11 to 35 percent. The CDS and FARS-based point estimates are both within these confidence bounds.

4.2 Effect on fatalities

If a fatally injured occupant's only injury had been a head injury due to upper-interior contact, preventing that injury would have saved the occupant's life. If someone's injuries had all been due to sources other than the upper interior, FMVSS No. 201 would have had no effect. But the situation is more complicated if the occupant had both types of injuries. Preventing the head injuries due to upper-interior contact could reduce fatality risk by varying extents, depending on the number and severity of the other injuries. NHTSA has developed complex models requiring detailed data to address the issue.[51]

This report uses a simpler approach that, although inaccurate for individual cases, may estimate the aggregate effect over the entire database. It is based on 2,305 fatally injured occupant cases on 1995-2009 CDS, who were seated at designated positions in pre-FMVSS No. 201 vehicles and had at least one reported AIS 3-to-6 injury with known contact source and body region. Moreover, as in Chapter 2, front-seat occupants in frontal crashes are excluded if there is no frontal air bag, outboard occupants in non-frontal crashes are excluded if there is a torso bag, and so are any seats equipped with 2-point automatic seat belts. The objective is to obtain a population of fatality cases representative of typical vehicles just before FMVSS No. 201.

These 2,305 fatally-injured people experienced a total of 11,604 reported AIS 3-to-6 injuries with known contact source and body region. The potential fatality-reducing efficacy of FMVSS No. 201 was gauged by a "fault-tree" analysis:

- If the occupant had any AIS 6 injury (e.g., decapitation) that was not a head injury due to upper-interior contact, FMVSS No. 201 would not potentially save the occupant's life.

---

[51] Martin, P. G., & Eppinger, R. E. (2005). A Method to Attribute Fatalities and Costs to Specific Injuries, *Proceedings of the Nineteenth International Technical Conference on the Enhanced Safety of Vehicles*. (Paper No. 05-0220-O). Washington, DC: National Highway Traffic Safety Administration. Available at www-nrd.nhtsa.dot.gov/pdf/esv19/05-0220-O.pdf.

- If the maximum AIS among the upper-interior head injuries is lower than the maximum AIS among the other injuries, FMVSS No. 201 is likewise assumed to have no potential fatality-reducing effectiveness.
- But if the maximum AIS among the upper-interior head injuries is higher than the maximum AIS among the other injuries, FMVSS No. 201 is assumed to have the potential to reduce fatality risk by 23.73 percent (the best estimate for AIS 4-to-6 reduction).
- If the maximum AIS among the upper-interior head injuries equals the maximum AIS among the other injuries:
    - If the number of maximum-severity upper-interior head injuries is less than the number of maximum-severity other injuries, FMVSS No. 201 is assumed to have no potential fatality-reducing effectiveness.
    - If the number of maximum-severity upper-interior head injuries is more than the number of maximum-severity other injuries, FMVSS No. 201 is assumed to reduce fatality risk by 23.73 percent.
    - If the number of maximum-severity upper-interior head injuries equals the number of maximum-severity other injuries, FMVSS No. 201 is assumed to reduce fatality risk by 11.865 percent (half of the best estimate).

When the occupant cases are weighted by RATWGT, FMVSS No. 201 had a potential 23.73 percent effect for 16.72 percent of the occupants and a potential 11.865 percent effect for 2.60 percent of the occupants.[52] The estimated fatality-reducing effectiveness of FMVSS No. 201 (without curtain bags) for car and LTV occupants at designated seating positions is:

$$.1672 \times .2373 + .0260 \times .5 \times .2373 = .1802 \times .2373 = 4.28 \text{ percent.}$$

The 95-percent confidence bounds for this estimate of fatality reduction range from 2.0 to 6.2 percent.

4.3 Lives saved and potentially savable by FMVSS No. 201 in CY 2004-2009

The number of lives potentially savable in a year by FMVSS No. 201 when every vehicle on the road meets the standard is the effectiveness estimate, 4.28 percent times the baseline number of annual fatalities to people seated at designated positions in cars or LTVs. The number of "baseline" fatalities in any recent year (specifically 2004... 2009) is the number of actual fatalities incremented by the lives that were already saved by FMVSS No. 201 or by two other safety technologies that have entered or shortly will enter the vehicle fleet, but in most cases after FMVSS No. 201:

- Curtain, torso, and/or combination air bags; and
- Electronic stability control.

In other words, some of the vehicles on the road were already furnished with some or all of the new safety equipment; the "baseline" is the higher, hypothetical number of fatalities that likely would have occurred if none of the vehicles had been so equipped. Lives saved by earlier

---

[52] With unweighted CDS data, the corresponding percentages are quite similar: 16.27% and 2.34%, respectively.

technologies such as frontal air bags do not need to be estimated, because the baseline is the typical vehicle just before FMVSS No. 201 went into effect (1999-2003) and this vehicle was already equipped with frontal air bags and many other earlier technologies.

The actual number of car- and LTV-occupant fatalities at designated seat positions declined greatly during 2004 to 2009, starting at 31,141 in 2004, tapering gradually to 30,937 in 2005, 30,021 in 2006, and 28,556 in 2007, and then plummeting to 24,986 in 2008 and 22,937 in 2009. A portion of the downward trend, as will be discussed, is due to growing availability of the recent technologies such as ESC; another portion is due to the retirement of older vehicles not yet equipped even with earlier safety technologies such as frontal air bags; but much of it, especially in 2008-2009 is due to changes in the number and type of VMT or other factors.

In 2004, NHTSA described a method that operates on FARS data to estimate lives saved by existing safety technologies.[53] In 2007, NHTSA adapted the method to estimate the baseline fatalities that would have occurred if no vehicles had been equipped with certain recently-developed safety technologies.[54] Each 100 actual fatality cases on FARS represent a theoretically even greater number of fatalities that could have happened if the vehicles had not been equipped with those technologies. The actual fatality cases are the starting point for a model to estimate the hypothetical baseline fatalities.

For example, FARS might have records of 100 occupant fatalities in rollover crashes of cars equipped of a particular make, model, and year equipped with ESC and certified to FMVSS No. 201, but not equipped with side or curtain air bags. The estimated effectiveness of ESC in preventing car rollovers is 56 percent.[55] If this cohort of cars (i.e., this particular make, model, and year) had not been equipped with ESC, there would have been not 100 but

$$100/(1 - .56) = 227 \text{ would-be fatalities.}$$

In other words, the existence of these 100 actual rollovers with ESC on FARS implies that, without ESC, this cohort of cars would have been involved in 227 fatal rollovers. However, 127 of those could-have-been fatal rollovers did not become FARS cases because ESC prevented the crash entirely or reduced its severity to a nonfatal event.

FMVSS No. 201 certification (without head-protection air bags) is associated with a 4.28-percent reduction of overall fatality risk. If, in addition to removing the ESC, the head protection in those cars had been degraded to pre-FMVSS No. 201 levels, the number of would-be fatalities would have increased from 227 to

$$227/(1 - .0428) = 237 \text{ would-be fatalities.}$$

---

[53] Kahane, C. J. (2004, October). *Lives saved by the Federal Motor Vehicle Safety Standards and other vehicle safety technologies, 1960-2002.* (Report. DOT HS 809 833, pp. 25-31). Washington, DC: National Highway Traffic Safety Administration. Available at www-nrd.nhtsa.dot.gov/Pubs/809833.PDF.
[54] Kahane (2007), Chapter 4.
[55] Sivinski, R. (2011, June). *Crash prevention effectiveness of light-vehicle electronic stability control: An update of the 2007 NHTSA evaluation.* (Report No. DOT HS 811 486). Washington, DC: National Highway Traffic Safety Administration. Available at www-nrd.nhtsa.dot.gov/Pubs/811486.pdf.

FMVSS No. 201 saved 237 – 227 = 10 lives. The 100 actual fatality cases on FARS imply the existence of 237 could-have-been fatal crash cases. ESC and FMVSS No. 201 certification saved a combined total of 137 lives: 127 by ESC and 10 by FMVSS No. 201.

The model begins with the actual FARS fatality cases and inflates them step-by-step as it "removes" three safety technologies from the vehicles, starting with the generally most recent (ESC) and ending with the earliest of the three (FMVSS No. 201). Of course, if a specific vehicle on FARS is not equipped with a particular technology, there is nothing to "remove" and nothing to inflate for that vehicle on that step. The model uses the agency's published effectiveness estimates for ESC[56] and side/curtain air bags[57] and the 4.28-percent reduction for FMVSS No. 201.[58] If the vehicle is certified to FMVSS No. 201 and equipped with side or curtain air bags and if the crash is a type of side impact where the bag has some effect, no additional benefit is counted for FMVSS No. 201.[59] At each step into the past, the model tallies the lives saved by the latest safety technology – i.e., the additional fatalities that would have occurred without it.

Table 4-1 shows the percent of cars and LTVs on the road certified to FMVSS No. 201, equipped with any type of curtain or side air bag, and equipped with ESC. The percentages grow relatively slowly because in every year there are large numbers of older vehicles, not equipped with any of those technologies, still on the road. By 2009, close to half the vehicles on the road were certified to FMVSS No. 201, but only 7 percent were equipped with ESC. It is also clear from Table 4-1 that, typically, FMVSS No. 201 certification preceded curtain or side air bags and ESC came last.

---

[56] Sivinski (2011): in first-event rollovers, 56% reduction in cars and 74% in LTVs; in other single-vehicle non-pedestrian crashes preceded by running off the road (e.g., impacts with fixed objects), 47% in cars and 45% in LTVs; in other single-vehicle crashes (e.g. occupant falling from moving vehicle), no effect; in all other crashes, 8%.
[57] Kahane (2007): in nearside impacts of cars and LTVs, 24% for curtain + torso or combination bags, 12% for curtain only or torso only; in far-side impacts of cars only, for unbelted occupants or for belted drivers without an accompanying front-seat passenger, 24% for curtain + torso, 12% for curtain only, no effect for torso only or combination bags.
[58] No effectiveness is assumed at this time for rear-seat curtains or for curtains that deploy in rollovers because the agency has not yet published evaluations of these technologies.
[59] To avoid double-counting, because Kahane (2007), pp. 14-16 and pp. 95-99 cannot exclude the possibility that some of the effect attributed to the air bags may actually derive from non-air-bag modifications to meet FMVSS No. 201, as these often coincided or nearly coincided with the installation of the air bags.

TABLE 4-1

PERCENT OF VEHICLES ON THE ROAD CERTIFIED TO FMVSS No. 201 OR
EQUIPPED WITH CURTAIN/SIDE AIR BAGS OR WITH ESC, CY 2004-2009
(FARS occupant fatality cases)

Calendar Year	FMVSS 201	Curtain, Torso and/or Combo Air Bags	ESC
2004	20.0	6.0	1.4
2005	25.9	7.5	2.0
2006	31.3	9.6	2.8
2007	36.5	12.0	3.8
2008	42.0	15.1	5.6
2009	45.3	17.7	7.0

Table 4-2 presents the results of the model, namely, the lives saved by each of the three technologies and the adjusted baseline fatalities, which are the actual fatalities on FARS plus the sum of the lives saved by the three technologies.

TABLE 4-2

ADJUSTED BASELINE FATALITIES: ACTUAL FATALITIES PLUS LIVES
SAVED BY THREE TECHNOLOGIES, CY 2004-2009
(Occupants of cars and LTVs at designated seating positions)

| CY | Lives Saved | | | | Fatalities | |
	FMVSS 201	Curtain/ Side Bags	ESC	Total	Actual	Adjusted Baseline
2004	270	107	225	602	31,141	31,743
2005	350	124	301	775	30,937	31,712
2006	410	170	407	988	30,021	31,009
2007	459	193	554	1,205	28,556	29,761
2008	465	222	723	1,410	24,986	26,396
2009	460	242	772	1,473	22,937	24,410
	2,415	1,057	2,982	6,453		

As the proportion of the on-road fleet certified to FMVSS No. 201 grew from 20 to 45 percent, life-savings increased from 270 in 2004 to 460 in 2009. Benefits likewise increased for curtain and side air bags and became substantial for ESC even though only a small percentage of vehicles on the road were equipped with it. The sum of lives saved grew from 602 in 2004 to 1,473 in 2009 and to some extent offset the large drop in actual fatalities. The adjusted baseline fatalities remained above 31,000 in 2004-2006 but had dropped to 24,410 by 2009.

Lives potentially savable. The potential annual fatality reduction for FMVSS No. 201 depends on the choice of base years. If CY 2004-2007 is selected as the base, the adjusted baseline fatalities – the number that would have occurred if no vehicles had certified to FMVSS No. 201 or been equipped with curtain/side air bags or ESC – averaged 31,056. If all of those vehicles had then certified to FMVSS No. 201, fatalities would have decreased by an estimated 4.28 percent, a savings of 1,329 lives per year. The 95-percent confidence bounds on effectiveness, ranging from 2.0 to 6.2 percent, indicate a range of 621 to 1,925 lives saved per year. If CY 2008-2009 is selected as the base, adjusted baseline fatalities averaged 25,403. With that lower baseline, the point estimate is 1,087 lives saved per year, with a confidence range from 500 to 1,575.

4.4 Cost of FMVSS No. 201; cost per life saved

In 2003, NHTSA contractors estimated the purchase-price increase and mass added to passenger cars and LTVs as a consequence of the changes made by the automotive industry to meet the FMVSS No. 201 upgrade without head-protection air bags.[60] The contractor studied pre-standard passenger vehicles of 10 makes and models and post-standard vehicles of the same or corresponding models. The analysis was based on "teardown" or "reverse engineering" – i.e., detailed inspection and disassembly of the components in the head-impact areas. The purchase-price increase includes the materials, labor, tooling, and variable burden needed to produce the parts and assemble them into the vehicle, plus mark-ups by the parts suppliers (where applicable), the auto manufacturers, and dealers.[61] In addition to providing cost estimates, the study describes in detail what energy-absorbing materials were actually added or modified, as summarized in Section 1.3. The vehicles comprise a variety of manufacturers and included six passenger cars, a pickup truck, an SUV and two minivans. In addition to the contractor's own examinations of the components, the contractor received detailed information from the manufacturers, such as parts lists, identifying how vehicles were modified.

Purchase-price and mass increases are estimated for the 10 models in Table 4-3. The purchase-price increase has been translated to 2010 dollars from the 2003 dollars in the contractor's report, based on the gross-domestic-product (GDP) deflator.[62]

---

[60] Ludtke, Osen, Gladstone, & Lieberman (2003).

[61] Tarbet, M. J. (2004, December). *Cost and weight added by the Federal Motor Vehicle Safety Standards for model years 1968-2001 in passenger cars and light trucks*. (Report No. DOT HS 809 834). Washington, DC: National Highway Traffic Safety Administration. Available at www-nrd nhtsa.dot.gov/Pubs/809834.PDF.

[62] BEA (2011). Current-Dollar and "Real" GDP. Washington, DC: Bureau of Economic Analysis. http://www.bea.gov/national/index.htm; in 2003, GDP was $11,142.1 billion in current dollars and $11,840.7 billion in 2005 dollars; in 2010, GDP was $14,660.4 billion in current dollars and $13,248.2 billion in 2005 dollars – i.e., the ratio of current to 2005 dollars increased by 17.6% from 2003 to 2010.

TABLE 4-3

COST AND WEIGHT ADDED BY FMVSS No. 201 TO 10 CARS AND LTVs
WITHOUT HEAD-PROTECTION AIR BAGS

Make and Model	Model Years Studied Pre-Std.	Post-Std.	Purchase-Price Increase (2010 $)	Mass Added (Pounds)
Passenger Cars				
Ford Crown Victoria	2000	2001	$22.06	1.27
Ford Taurus	1998	2001	24.54	1.07
Volkswagen Jetta	1998	2002	12.71	1.95
Honda Accord	1998	2001	3.43	.02
Toyota Camry	1998	2002	.61	.10
Kia Spectra	2000	2002	6.12	.78
Average			11.58	.87
Excluding Camry			13.77	1.02
Sales-weighted average			11.52	.66
Excluding Camry			14.92	.84
LTVs				
Jeep Grand Cherokee	1998	1999	23.64	4.91
Dodge Caravan	2000	2001	36.61	8.49
Ford F-150 Pickup	1999	2000	6.06	1.32
Pontiac Montana	1998	2001	13.45	5.40
Average			19.94	5.03
Sales-weighted average			20.33	4.89
Cars and LTVs				
Average			14.92	2.53
Excluding Camry			16.51	2.80
Sales-weighted average			15.29	2.47
Excluding Camry			17.60	2.84

The large model-to-model variations in cost and weight probably reflect, above all, the extent to which vehicles had to be modified to certify to FMVSS No. 201. Some pre-standard vehicles required little or no change in some of their head-impact zones to meet FMVSS No. 201. The variations also reflect different materials used in the vehicles as described in Section 1.3. The cost for Toyota Camry was lower than any other car, perhaps because the 1997 and 1998 models, although nominally pre-standard, may have already incorporated some of the materials needed to meet the impending requirement (the Camry was redesigned in 1997). This possibility has been acknowledged in the statistical analyses of Chapters 2 and 3, where only 1996 and earlier models of Camry are included in the "pre-standard" cohort and 1997-1998 models excluded from the analyses. Excluding Camry from the cost computations probably gives a more meaningful estimate of the cost and weight added by FMVSS No. 201 and one that is more consistent with the statistical analyses.

For the remaining five models of passenger cars, the arithmetic average is a purchase-price increase of $13.77 and a mass increase of 0.87 pounds. The sales-weighted[63] averages are $14.92 and 0.84 pounds. There is little difference between the arithmetic and sales-weighted averages.

For the four LTVs, the arithmetic average is $19.94 and 5.03 pounds. Sales-weighted averages are $20.33 and 4.89 pounds. The cost and weight added by FMVSS No. 201 is higher in LTVs than in cars largely because LTVs needed more improvements to their various target areas to meet FMVSS No. 201. As discussed in Section 1.4, HIC test results improved by an average of 172 units per target area in the post-Standard versus pre-Standard tests of passenger cars, and by 301 units per target area in LTVs.

For the five cars (excluding Camry) and four LTVs combined, the arithmetic average price increase is $16.51 and the mass increase 2.80 pounds. The sales-weighted averages are $17.60 and 2.84 pounds.

Total cost. The sales-weighted averages for nine makes and models (excluding Toyota Camry) will serve as the best estimate of the purchase-price and mass increases per vehicle for FMVSS No. 201 without head-protection air bags: $17.60 in 2010 dollars and 2.84 pounds.

The increase of 2.84 pounds will add to the vehicle's lifetime fuel consumption. NHTSA's 2009 final regulatory impact analysis of FMVSS No. 216 (roof crush strength) estimates that each added pound of weight increases lifetime fuel consumption by 1.356 gallons. The additional fuel is not purchased all at once but gradually over the life of the vehicle. At the 2010 average price retail price of gasoline, $2.79 per gallon, the net present value (NPV) of the lifetime 1.356-gallon increase per added pound is also $2.79.[64] The total cost of FMVSS No. 201 per vehicle is:

---

[63] Based on total sales in MY 1999-2003, including sales of closely related models: Mercury Grand Marquis with Ford Crown Victoria; Mercury Sable with Ford Taurus; Kia Sephia with Kia Spectra; Dodge Grand Caravan, Plymouth Voyager/Grand Voyager, and Chrysler Town & Country with Dodge Caravan; all cab types of Ford F-150; and Chevrolet Venture and Oldsmobile Silhouette with Pontiac Montana. "Sales" are maximum registrations in CY 2000-2005 on R.L. Polk's National Vehicle Population Profile.

[64] NHTSA (2009). *Final Regulatory Impact Analysis, FMVSS No. 216 Upgrade, Roof Crush Resistance.* Docket No. NHTSA-2009-0093-4. Washington, DC: National Highway Traffic Safety Administration, p. 103; the calculation, based on 2007 fuel prices and excluding fuel taxes in computing the NPV, is that a 13.27-pound increase results in consumption of 18 gallons and has NPV $37.35; the price of fuel dropped from $2.81 to $2.79 per gallon from 2007 to 2010, based on the yearly average of the Energy Information Administration's 52 weekly estimates for "all grades, entire country," EIA (2011). Retail Gasoline Historical Prices. Washington, DC: Energy Information Administration. http://www.eia.gov/oil_gas/petroleum/data_publications/wrgp/mogas_history.html.

$$17.60 + 2.84 \times \$2.79 = \$25.52 \text{ in 2010 dollars}$$

Car plus LTV sales in the United States averaged 16,602,000 per year from 2004 to 2007.[65] At $25.52 per vehicle, that amounts to an annual cost of $424 million for FMVSS No. 201 without head-protection air bags, in 2010 dollars. In 2008-2009, sales decreased to an average of 11,799,000. The annual cost of FMVSS No. 201 would have been $301 million.

Cost per life saved.[66] The cost per life saved is based on a "steady-state" calculation: the annual cost of certifying each new vehicle to FMVSS No. 201 divided by the potential annual fatality reduction when all vehicles on the road meet FMVSS No. 201 relative to when none of them meet FMVSS No. 201. The potential annual fatality reduction for FMVSS No. 201 was estimated in Section 4.3 and it depended on the choice of base years. If CY 2004-2007 is selected as the base, the point estimate was 1,329 lives saved per year, with 95-percent confidence bounds ranging from 621 to 1,925 lives saved per year. At an annual cost of $424 million for these base years, the cost per life saved is $319,000, with confidence bounds ranging from $220,000 to $683,000.

If CY 2008-2009 is selected as the base, there are fewer baseline fatalities and, as a result, a lower estimate of potential annual fatality reduction 1,087 lives saved per year, with a confidence range from 500 to 1,575. But costs are also lower because of reduced sales: $301 million. The cost per life saved is $277,000, with confidence bounds ranging from $191,000 to $602,000.

If vehicle sales should return to the 2004-2007 levels while baseline fatalities continue, hopefully, to stay near the 2008-2009 levels, the cost per life saved would increase somewhat to $390,000, with confidence bounds ranging from $269,000 to $848,000.

All of these estimates, even the upper confidence bounds for cost per life saved show that FMVSS No. 201 is a highly cost-effective technology. The cost per life saved in all cases is far less than the Department of Transportation's estimate of $6,000,000-6,300,000 (in 2008 dollars) for the value of a statistical life (VSL).[67]

4.5 Upper-interior head injuries after FMVSS No. 201/head-protection air bags

Tables 1-2 and 1-3 in Section 1.1 showed how AIS 3-to-6 and AIS 4-to-6 head injuries due to upper-interior contact were distributed by contact source and injury type before FMVSS No. 201 and head-protection air bags. Tables 4-4 and 4-5 conclude the report by demonstrating how these distributions changed after FMVSS No. 201 and head-protection air bags – i.e., what types of contacts and injuries remain. In each table, the pre-FMVSS No. 201 distribution, copied from Section 1.1, is shown in italic type for comparison.

---

[65] *Ward's Automotive Yearbook 2011*. Southfield, MI: Ward's Automotive Group, p. 105.
[66] In regulatory evaluations or regulatory impact analyses, NHTSA frequently estimates the cost per equivalent life saved. Equivalent lives include fatalities and injuries valued at their relative comprehensive costs by AIS level compared to the comprehensive cost of a fatality. For this evaluation, we did not estimate injuries reduced by AIS level, thus the comparison has been limited to costs per life saved.
[67] NHTSA (2011). *Final Regulatory Evaluation, FMVSS No. 218 Motorcycle Helmet Labeling*. Docket No. NHTSA-2011-0050-0002. Washington, DC: National Highway Traffic Safety Administration. Appendix C.

Table 4-4 reveals that the distribution of injuries by contact source has not changed that much (even though the risk of AIS 4-to-6 injuries has been significantly reduced). The B-pillar and front header now account for a somewhat larger proportion of the injuries than they did before FMVSS No. 201, whereas the A-pillar and roof interior, a smaller share. That happened because FMVSS No. 201 is especially effective in reducing injuries from the A-pillar and the roof interior.

TABLE 4-4

UPPER-INTERIOR SOURCES OF HEAD INJURIES AFTER FMVSS No. 201
AND HEAD-PROTECTION AIR BAGS
(Unweighted 1995-2009 CDS)

	AIS 3 to 6		AIS 4 to 6	
	N	%	N	%
Front header, sun visor	36	11.2	16	12.1
Windshield + A-pillar and/or front header	16	5.0	8	6.1
A-pillar	47	14.6	16	12.1
B-pillar	77	23.9	39	29.6
Other pillar	6	1.9	3	2.3
Roof side rail	60	18.6	18	13.6
Side window + pillars and/or roof side rail	none	.	none	.
Rear header	4	1.2	1	.8
Roof interior, maplite, sunroof	76	23.6	31	23.5
	322	100.0	132	100.0

Before FMVSS No. 201 and Without Head-Protection Air Bags

Front header, sun visor	*425*	*10.4*	*190*	*9.5*
Windshield + A-pillar and/or front header	*215*	*5.1*	*111*	*5.6*
A-pillar	*719*	*17.0*	*307*	*15.4*
B-pillar	*971*	*22.9*	*445*	*22.3*
Other pillar	*120*	*2.8*	*48*	*2.4*
Roof side rail	*614*	*14.5*	*300*	*15.0*
Side window + pillars and/or roof side rail	*24*	*.6*	*13*	*.7*
Rear header	*44*	*1.0*	*18*	*.9*
Roof interior, maplite, sunroof	*1,100*	*26.0*	*563*	*28.2*
	4,232	*100.0*	*1,995*	*100.0*

Table 4-5 shows a great reduction in concussions after FMVSS No. 201 and head-protection air bags, offset by a relative increase in "other head injuries." The remaining injury types neither gained nor lost share. About 75 percent of the "other head injuries" are brain lacerations. Table 4-5 does not imply that the risk of brain lacerations has increased; rather, because FMVSS No.

201 has been so effective against concussions and some other types of brain injury, the lacerations now account for a larger share of the injuries that still remain.

TABLE 4-5

UPPER-INTERIOR HEAD INJURIES AFTER FMVSS No. 201
AND HEAD-PROTECTION AIR BAGS, BY INJURY TYPE
(Unweighted 1995-2009 CDS)

	AIS 3 to 6		AIS 4 to 6	
	N	%	N	%
Concussion	2	.6	1	.8
Brain contusion	41	12.7	7	5.3
Unknown brain injury	188	58.4	80	60.6
Skull fracture	48	14.9	15	11.4
Facial fractures	10	3.1	none	.
All other head injuries	33	10.3	29	22.0
	322	100.0	132	100.0

Before FMVSS No. 201 and Without Head-Protection Air Bags

Concussion	*346*	*8.2*	*202*	*10.1*
Brain contusion	*536*	*12.7*	*116*	*5.8*
Unknown brain injury	*2,388*	*56.4*	*1,210*	*60.7*
Skull fracture	*614*	*14.5*	*227*	*11.4*
Facial fractures	*100*	*2.4*	*none*	*.*
All other head injuries	*248*	*5.9*	*240*	*12.0*
	4,232	*100.0*	*1,995*	*100.0*

APPENDIX A

INITIAL MODEL YEAR OF FMVSS No. 201 CERTIFICATION
MAKES AND MODELS PRODUCED IN 1999-2003

MM2 codes[68]	Make and Model	Initial FMVSS No. 201 Certification
2300-2303	Jeep Cherokee	never (produced through 2001)
2312-2313	Jeep Grand Cherokee	1999 (mid-model-year)
2321	Jeep Wrangler	2003
2342-2343	Jeep Liberty	always (produced starting 2002)[69]
6041-6042	Chrysler Concorde & LHS	1999 (mid-model-year)
6043	Chrysler Sebring convertible	2002
6043	Chrysler Sebring coupe	2002
6043	Chrysler Sebring sedan	2001
6044	Chrysler Cirrus	2001
6051	Chrysler 300	some in 1999 and 2000, all in 2001
6052	Chrysler PT Cruiser, SUV only	2003
6400-6409	Chrysler Town & Country	2002[70]
7013	Dodge Viper	2003
7020	Dodge Neon	some in 2000 and 2001, all in 2002
7041	Dodge Intrepid	some in 1999 and 2000, all in 2001
7042	Dodge Avenger	2002
7043	Dodge Stratus coupe	2002
7043	Dodge Stratus sedan	2001
7200-7205	Dodge Dakota	2000
7210-7215	Dodge Ram 1500 pickup	2002
7220-7235	Dodge Ram 2500/3500 pickup	2003
7312-7313	Dodge Durango	2001
7400-7409	Dodge Caravan & Grand Caravan	2002[71]
7410-7439	Dodge Ram Van	2003
9020	Plymouth Neon	some in 2000 and 2001, all in 2002
9038	Plymouth Breeze	never (produced through 2000)
9400-9409	Plym Voyager & Grand Voyager	never (produced through 2000)
12003	Ford Mustang convertible	2001
	Ford Mustang coupe	2003
12004	Ford Thunderbird	2002
12013	Ford Escort	2000
12016	Ford Crown Victoria	2001

---

[68] The 5-digit make-model code generated by the VIN analysis programs developed by NHTSA staff for use in evaluations of FMVSS and available to the public at www.nhtsa.gov/fuel-economy.

[69] FMVSS No. 201 certification began 9/1/2001 but production began 8/1/2001; probably reasonable to assume all Jeep Liberty met FMVSS No. 201, because unlikely to have been modified after a month.

[70] Not certified in 2001 but may have already met FMVSS No. 201 because redesigned one year before certification.

[71] Not certified in 2001 but may have already met FMVSS No. 201 because redesigned one year before certification.

MM2 codes	Make and Model	Initial FMVSS No. 201 Certification
12017	Ford Taurus	2000
12037	Ford Focus	2000
12200-12205	Ford Ranger	2003
12210-12219	Ford F-150 pickup	2002
12220-12239	Ford F-250/350 GVWR < 10,000	2003
12300-12309	Ford Explorer all types	2002
12312-12313	Ford Expedition	2003
12332-12333	Ford Excursion	2003
12342-12343	Ford Escape	2001
12400-12402	Ford Windstar	1999
12410-12439	Ford van GVWR < 10,000	2002
13001	Lincoln Town Car	2001
13005	Lincoln Continental	2002
13012	Lincoln LS	2000
13302-13309	Lincoln Aviator	2002
13312-13313	Lincoln Navigator	2003
14016	Mercury Grand Marquis	2001
14017	Mercury Sable	2000
14038	Mercury Cougar	2002
14302-14309	Mercury Mountaineer	2002
14450,14452	Mercury Villager	1999
18002	Buick LeSabre	2000
18003	Buick Park Avenue	2003
18017	Buick Century	2003
18020	Buick Regal	2003
18356,18357	Buick Rendezvous	always (produced starting 2002)
19003	Cadillac DeVille	2000
19014	Cadillac Seville	2002
19312-19343	Cadillac Escalade	2002
20002	Chevrolet Impala	always (produced starting 2000)
20004	Chevrolet Corvette	2003
20016	Chevrolet Cavalier	2003
20032	Chevrolet Prizm	2003
20036	Chevrolet Monte Carlo	2000
20037	Chevrolet Malibu	2003
20200-20205	Chevrolet S/T pickup	2003
20210-20215	Chevrolet C/K pickup	never
	Chevrolet Silverado 1500	always
20300-20303	Chevrolet Blazer[72]	2003
20302-20307	Chevrolet Trailblazer	always (produced starting 2002)
20312-20327	Chevrolet Tahoe, Suburban	2000
20330-20333	Chevrolet Tracker	1999
20400-20407	Chevrolet Astro van	2003
20410-20439	Chevrolet full-size van	2003
20450-20457	Chevrolet Venture	2002

---

[72] Excluding Trailblazer

MM2 codes	Make and Model	Initial FMVSS No. 201 Certification
21021	Oldsmobile Alero	1999
21022	Oldsmobile Aurora	2001
21302-21303	Oldsmobile Bravada	2002
21450-21457	Oldsmobile Silhouette	2002
22002	Pontiac Bonneville	2000
22016	Pontiac Sunfire	2003
22018	Pontiac Grand Am	1999
22020	Pontiac Grand Prix	2003
22352-22353	Pontiac Aztek	always (produced starting 2001)
22450-22457	Pontiac Montana	2002
23200-23205	GMC Sonoma pickup	2003
23210-23215	GMC Sierra (like C/K)	never
	GMC Sierra (like Silverado)	always
23300-23303	GMC Jimmy	never
23302-23307	GMC Envoy	always (produced starting 2002)
23312-23328	GMC Yukon	2000
23400-23407	GMC Safari van	2003
23410-23439	GMC full-size van	2003
24005-24006	Saturn L	2003
24362-24363	Saturn Vue	always (produced starting 2002)
30040	Volkswagen Jetta	upon redesign to CG 30010 in 1999
30042	Volkswagen Golf	2002
	Volkswagen GTI	2003
30046	Volkswagen Passat	2002 (except possibly W8 model)
30047	Volkswagen New Beetle	2002
32...	Audi	certified with head-protection air bags
33...	Mini-Cooper	certified with head-protection air bags
34...	BMW	certified with head-protection air bags
35039	Nissan Maxima	2000
35043	Nissan Sentra	2000
35047	Nissan Altima	1999
35200-35205	Nissan Frontier	2002
35302-35303	Nissan Pathfinder	2003
35322-35323	Nissan Xterra	always (produced starting 2000)
35452	Nissan Quest	1999
37031	Honda Civic	2001
37032	Honda Accord	1999
37035,37037	Honda S2000, Insight	always (produced starting 2000)
37302-37303	Honda CR-V	2002
37322-37323	Honda Passport	1999
37402	Honda Odyssey	1999
38300,38301	Isuzu Rodeo Sport	2003
38302-38307	Isuzu Ascender	always (produced starting 2002)
38322-38323	Isuzu Rodeo	1999
38326-38327	Isuzu Axiom	always (produced starting 2002)
39031,39032	Jaguar XK, XJ	2003

MM2 codes	Make and Model	Initial FMVSS No. 201 Certification
39034	Jaguar S-Type	2000
39036	Jaguar X-Type	always (produced starting 2002)
41035	Mazda Protégé	1999 (except with sliding roofs)
41037	Mazda 626	never (produced through 2002)
41045	Mazda Miata	2003
41047	Mazda Millenia	never (produced through 2002)
41050	Mazda 6	always (produced starting 2003)
41200-41205	Mazda B-Pickup	2003
41342-41343	Mazda Tribute	2001
41402	Mazda MPV	2000
42042	Mercedes C	2001
42043,42046	Mercedes S, CL	2000
42044	Mercedes SL	2003
42045	Mercedes SLK	2002
42047	Mercedes CLK	2003
42048	Mercedes E	2000
42303,42307	Mercedes ML	2000
45031,45040	Porsche	2000
47...	Saab	certified with head-protection air bags
48034	Subaru Legacy	2000 w/o sun roofs, 2001 with sun roofs
48038	Subaru Impreza	2002
48303	Subaru Forester	unknown (2002 or 2003)
49032	Toyota Corolla	2003
49033	Toyota Celica	2000
49040	Toyota Camry (excl Solara)	1999[73]
49041	Toyota MR2	2000
49043	Toyota Avalon	2003
49044	Toyota Camry Solara	2000
49045	Toyota Echo	2002
49046	Toyota Prius	2003
49200-49205	Toyota Tacoma	2002
49210-49215	Toyota Tundra	2003
49302-49303	Toyota 4Runner	2003
49313	Toyota Landcruiser	2003
49322-49323	Toyota RAV4	2001
49342-49343	Toyota Highlander	always (produced starting 2001)
49352-49353	Toyota Sequoia	always (produced starting 2001)
49402-49403	Toyota Sienna	2001
51...	Volvo	certified with head-protection air bags
52034	Mitsubishi Galant	1999 w/o sun roofs, 2002 with sun roofs
52037	Mitsubishi Eclipse	2000
52040	Mitsubishi Diamante	2003
52046	Mitsubishi Lancer	always (produced starting 2002)
52333	Mitsubishi Montero	2001

---

[73] 1997-1998 Camry may have been designed to meet some FMVSS No. 201 requirements; additional improvements in 2002.

MM2 codes	Make and Model	Initial FMVSS No. 201 Certification
52336-52337	Mitsubishi Montero Sport	2003
53032	Suzuki Esteem	never (produced through 2002)
53033	Suzuki Aerio	always (produced starting 2002)
53310-53313	Suzuki Sidekick	1999
53330-53337	Suzuki Vitara, Grand Vitara	1999
53342-53343	Suzuki XL-7	always (produced starting 2001)
54033	Acura NSX	2002
54035	Acura TL	1999
54036	Acura RL	2003
54037	Acura CL	2001
54038	Acura RSX	always (produced starting 2002)
54323	Acura MDX	always (produced starting 2001)
55033	Hyundai Sonata	1999
55035	Hyundai Elantra	2001
55036	Hyundai Accent 3-door hatchback	2000
	Hyundai Accent 4-door sedan	unknown
55037	Hyundai Tiburon	2002
55038	Hyundai XG350	2001 w/o sun roof, 2003 with sun roof
55302-55309	Hyundai Santa Fe	always (produced starting 2001)
58032	Infiniti Q45	certified with head-protection air bags
58035	Infiniti I30	certified with head-protection air bags
59031	Lexus ES	1999[74]
59032	Lexus LS	certified with head-protection air bags
59033	Lexus SC	2001
59034	Lexus GS	1999[75]
59035	Lexus IS	certified with head-protection air bags
59313	Lexus LX470	2003
59332-59343	Lexus RX	2003
62303	Land Rover Discovery	never (produced through 1999)
62307	Land Rover Discovery II	always (produced starting 1999)
62313	Land Rover Range Rover	certified with head-protection air bags
62341, 62343	Land Rover Freelander	always (produced starting 2002)
63031	Kia Sephia	1999[76]
63032	Kia Rio	always (produced starting 2001)
63033	Kia Spectra	2001 for sure, 2000 not sure but likely
63034	Kia Optima	always (produced starting 2001)
63300-36603	Kia Sportage	2002
63402	Kia Sedona	always (produced starting 2002)
64031	Daewoo Lanos	2002
64032	Daewoo Nubira	2001
64033	Daewoo Leganza	1999

---

[74] 1997-1998 ES may have been designed to meet some FMVSS No. 201 requirements; additional improvements in 2002.
[75] 1998, the first year of a redesigned Lexus GS, may have already anticipated some parts of FMVSS No. 201.
[76] 1998, the first year of a redesigned Sephia, may have already anticipated some parts of FMVSS No. 201.

APPENDIX B

LISTING AND CLASSIFICATION OF ICD-10 S AND T CODES

Red = head injuries, excluding minor injuries
Blue = control group injuries (excluding minor injuries)
Black = excluded from analysis (minor injuries, neck injuries, multiple/unspecified body regions)

REGION	SEVERITY	CODE	INJURY
Head	Minor	S00.0	Superficial injury of scalp
Head	Minor	S00.3	Superficial injury of nose
Head	Minor	S00.5	Superficial injury of lip and oral cavity
Head	Minor	S00.7	Multiple superficial injuries of head
Head	Minor	S00.8	Superficial injury of other parts of head
Head	Minor	S00.9	Superficial injury of head, part unspecified
Head	Minor	S01.0	Open wound of scalp
Head	Minor	S01.1	Open wound of eyelid and periocular area
Head	Minor	S01.2	Open wound of nose
Head	Minor	S01.3	Open wound of ear
Head	Minor	S01.5	Open wound of lip and oral cavity
Head	Minor	S01.7	Multiple open wounds of head
Head	Minor	S01.8	Open wound of other parts of head
Head	Minor	S01.9	Open wound of head, part unspecified
Head	Severe	S02.0	Fracture of vault of skull
Head	Severe	S02.1	Fracture of base of skull
Head	Minor	S02.2	Fracture of nasal bones
Head	Severe	S02.3	Fracture of orbital floor
Head	Severe	S02.4	Fracture of malar and maxillary bones
Head	Minor	S02.5	Fracture of tooth
Head	Severe	S02.6	Fracture of mandible
Head	Severe	S02.7	Multiple fractures involving skull and facial bones
Head	Unk Sev	S02.8	Fractures of other skull and facial bones
Head	Unk Sev	S02.9	Fracture of skull and facial bones, part unspecified
Head	Minor	S03.1	Dislocation of septal cartilage of nose
Head	Unk Sev	S03.2	Dislocation of other and unspecified parts of head
Head	Severe	S04.8	Injury of other cranial nerves
Head	Severe	S04.9	Injury of unspecified cranial nerve
Head	Severe	S05.1	Contusion of eyeball and orbital tissues
Head	Severe	S05.3	Ocular laceration without prolapse or loss of intraocular tissue
Head	Severe	S05.4	Penetrating wound of orbit with or without foreign body
Head	Severe	S05.7	Avulsion of eye
Head	Unk Sev	S05.8	Other injuries of eye and orbit
Head	Unk Sev	S05.9	Injury of eye and orbit, unspecified
Head	Severe	S06.0	Concussion
Head	Severe	S06.1	Traumatic cerebral edema
Head	Severe	S06.2	Diffuse brain injury
Head	Severe	S06.3	Focal brain injury
Head	Severe	S06.4	Epidural hemorrhage
Head	Severe	S06.5	Traumatic subdural hemorrhage
Head	Severe	S06.6	Traumatic subarachnoid hemorrhage
Head	Severe	S06.7	Intracranial injury with prolonged coma
Head	Severe	S06.8	Other intracranial injuries
Head	Severe	S06.9	Intracranial injury, unspecified
Head	Severe	S07.0	Crushing injury of face
Head	Severe	S07.1	Crushing injury of skull
Head	Severe	S07.3	Crushing injury of other parts of head
Head	Severe	S07.9	Crushing injury of head, part unspecified
Head	Severe	S08.0	Avulsion of scalp
Head	Severe	S08.8	Traumatic amputation of other parts of head
Head	Severe	S08.9	Traumatic amputation of unspecified part of head
Head	Severe	S09.0	Injury of blood vessels of head, not elsewhere classified
Head	Unk Sev	S09.1	Injury of muscle and tendon of head

REGION	SEVERITY	CODE	INJURY
Head	Severe	S09.2	Traumatic rupture of ear drum
Head	Unk Sev	S09.7	Multiple injuries of head
Head	Unk Sev	S09.8	Other specified injuries of head
Head	Unk Sev	S09.9	Unspecified injury of head
Neck	Minor	S10.0	Contusion of throat
Neck	Minor	S10.1	Other and unspecified superficial injuries of throat
Neck	Minor	S10.8	Superficial injury of other parts of neck
Neck	Minor	S10.9	Superficial injury of neck, part unspecified
Neck	Severe	S11.0	Open wound involving larynx and trachea
Neck	Severe	S11.7	Multiple open wounds of neck
Neck	Severe	S11.8	Open wound of other parts of neck
Neck	Severe	S11.9	Open wound of neck, part unspecified
Neck	Severe	S12.0	Fracture of first cervical vertebra (atlas)
Neck	Severe	S12.1	Fracture of second cervical vertebra (axis)
Neck	Severe	S12.2	Fracture of other specified cervical vertebra
Neck	Severe	S12.7	Multiple fractures of cervical spine
Neck	Severe	S12.8	Fracture of other parts of neck
Neck	Severe	S12.9	Fracture of neck, part unspecified
Neck	Severe	S13.0	Traumatic rupture of cervical intervertebral disc
Neck	Severe	S13.1	Dislocation of cervical vertebra
Neck	Severe	S13.3	Multiple dislocations of neck
Neck	Severe	S13.4	Sprain and strain of cervical spine
Neck	Severe	S13.6	Sprain and strain of joints and ligaments of o. & u. parts of neck
Neck	Severe	S14.0	Concussion and edema of cervical spinal cord
Neck	Severe	S14.1	Other and unspecified injuries of cervical spinal cord
Neck	Severe	S14.2	Injury of nerve root of cervical spine
Neck	Severe	S14.3	Injury of brachial plexus
Neck	Severe	S14.6	Injury of other and unspecified nerves of neck
Neck	Severe	S15.0	Injury of carotid artery
Neck	Severe	S15.1	Injury of vertebral artery
Neck	Severe	S15.2	Injury of external jugular vein
Neck	Severe	S15.3	Injury of internal jugular vein
Neck	Severe	S15.7	Injury of multiple blood vessels at neck level
Neck	Severe	S15.9	Injury of unspecified blood vessel at neck level
Neck	Minor	S16	Injury of muscle and tendon at neck level
Neck	Severe	S17.0	Crushing injury of larynx and trachea
Neck	Severe	S17.8	Crushing injury of other parts of neck
Neck	Severe	S17.9	Crushing injury of neck, part unspecified
Neck	Unk Sev	S18	Other and unspecified injuries of neck
Neck	Unk Sev	S19.7	Multiple injuries of neck
Neck	Unk Sev	S19.8	Other specified injuries of neck
Neck	Unk Sev	S19.9	Unspecified injury of neck
Torso	Minor	S20.1	Other and unspecified superficial injuries of breast
Torso	Minor	S20.2	Contusion of thorax
Torso	Minor	S20.3	Other superficial injuries of front wall of thorax
Torso	Minor	S20.7	Multiple superficial injuries of thorax
Torso	Minor	S20.8	Superficial injury of other and unspecified parts of thorax
Torso	Minor	S21.0	Open wound of breast
Torso	Minor	S21.1	Open wound of front wall of thorax
Torso	Minor	S21.2	Open wound of back wall of thorax
Torso	Minor	S21.7	Multiple open wounds of thoracic wall
Torso	Minor	S21.8	Open wound of other parts of thorax
Torso	Minor	S21.9	Open wound of thorax, part unspecified
Torso	Severe	S22.0	Fracture of thoracic vertebra
Torso	Severe	S22.1	Multiple fractures of thoracic spine
Torso	Severe	S22.2	Fracture of sternum
Torso	Severe	S22.3	Fracture of rib
Torso	Severe	S22.4	Multiple fractures of ribs
Torso	Severe	S22.5	Flail chest
Torso	Severe	S22.8	Fracture of other parts of bony thorax
Torso	Severe	S22.9	Fracture of bony thorax, part unspecified
Torso	Severe	S23.1	Dislocation of thoracic vertebra
Torso	Severe	S23.2	Dislocation of other and unspecified parts of thorax
Torso	Severe	S24.0	Concussion and edema of thoracic spinal cord
Torso	Severe	S24.1	Other and unspecified injuries of thoracic spinal cord
Torso	Severe	S24.2	Injury of nerve root of thoracic spine
Torso	Severe	S24.4	Injury of thoracic sympathetic nerves
Torso	Severe	S24.6	Injury of unspecified nerve of thorax

REGION	SEVERITY	CODE	INJURY
Torso	Severe	S25.0	Injury of thoracic aorta
Torso	Severe	S25.1	Injury of innominate or subclavian artery
Torso	Severe	S25.2	Injury of superior vena cava
Torso	Severe	S25.3	Injury of innominate or subclavian vein
Torso	Severe	S25.4	Injury of pulmonary blood vessels
Torso	Severe	S25.5	Injury of intercostal blood vessels
Torso	Severe	S25.7	Injury of multiple blood vessels of thorax
Torso	Severe	S25.8	Injury of other blood vessels of thorax
Torso	Severe	S25.9	Injury of unspecified blood vessel of thorax
Torso	Severe	S26.0	Injury of heart with hemopericardium
Torso	Severe	S26.8	Other injuries of heart
Torso	Severe	S26.9	Injury of heart, unspecified
Torso	Severe	S27.0	Traumatic pneumothorax
Torso	Severe	S27.1	Traumatic hemothorax
Torso	Severe	S27.2	Traumatic hemopneumothorax
Torso	Severe	S27.3	Other injuries of lung
Torso	Severe	S27.4	Injury of bronchus
Torso	Severe	S27.6	Injury of pleura
Torso	Severe	S27.7	Multiple injuries of intrathoracic organs
Torso	Severe	S27.8	Injury of other specified intrathoracic organs
Torso	Severe	S27.9	Injury of unspecified intrathoracic organ
Torso	Severe	S28.0	Crushed chest
Torso	Severe	S28.1	Traumatic amputation of part of thorax
Torso	**Minor**	**S29.0**	**Injury of muscle and tendon at thorax level**
Torso	Unk Sev	S29.7	Multiple injuries of thorax
Torso	Unk Sev	S29.8	Other specified injuries of thorax
Torso	Unk Sev	S29.9	Unspecified injury of thorax
Torso	Minor	S30.0	Contusion of lower back and pelvis
Torso	Minor	S30.1	Contusion of abdominal wall
Torso	Minor	S30.2	Contusion of external genital organs
Torso	Minor	S30.7	Multiple superficial injuries of abdomen, lower back and pelvis
Torso	Minor	S30.8	Other superficial injuries of abdomen, lower back and pelvis
Torso	Minor	S30.9	Superficial inj of abdomen, lower back and pelvis, part unspecified
Torso	Minor	S31.0	Open wound of lower back and pelvis
Torso	Minor	S31.1	Open wound of abdominal wall
Torso	Severe	S31.3	Open wound of scrotum and testes
Torso	Severe	S31.4	Open wound of vagina and vulva
Torso	**Minor**	**S31.7**	**Multiple open wounds of abdomen, lower back and pelvis**
Torso	**Minor**	**S31.8**	**Open wound of other and unspecified parts of abdomen**
Torso	Severe	S32.0	Fracture of lumbar vertebra
Torso	Severe	S32.1	Fracture of sacrum
Torso	Severe	S32.2	Fracture of coccyx
Torso	Severe	S32.3	Fracture of ilium
Torso	Severe	S32.4	Fracture of acetabulum
Torso	Severe	S32.5	Fracture of pubis
Torso	Severe	S32.7	Multiple fractures of lumbar spine and pelvis
Torso	Severe	S32.8	Fracture of other and unspecified parts of lumbar spine and pelvis
Torso	Severe	S33.1	Dislocation of lumbar vertebra
Torso	Severe	S34.1	Other injury of lumbar spinal cord
Torso	Severe	S34.8	Injury of o & u nerves at abdomen, lower back and pelvis level
Torso	Severe	S35.0	Injury of abdominal aorta
Torso	Severe	S35.1	Injury of inferior vena cava
Torso	Severe	S35.2	Injury of coeliac or mesenteric artery
Torso	Severe	S35.3	Injury of portal or splenic vein
Torso	Severe	S35.4	Injury of renal blood vessels
Torso	Severe	S35.5	Injury of iliac blood vessels
Torso	Severe	S35.8	Injury of other blood vessels at abdomen, lower back and pelvis level
Torso	Severe	S35.9	Injury of u blood vessel at abdomen, lower back and pelvis level
Torso	Severe	S36.0	Injury of spleen
Torso	Severe	S36.1	Injury of liver or gall bladder
Torso	Severe	S36.2	Injury of pancreas
Torso	Severe	S36.3	Injury of stomach
Torso	Severe	S36.4	Injury of small intestine
Torso	Severe	S36.5	Injury of colon
Torso	Severe	S36.6	Injury of rectum
Torso	Severe	S36.7	Injury of multiple intra-abdominal organs
Torso	Severe	S36.8	Injury of other intra-abdominal organs
Torso	Severe	S36.9	Injury of unspecified intra-abdominal organ

REGION	SEVERITY	CODE	INJURY
Torso	Severe	S37.0	Injury of kidney
Torso	Severe	S37.2	Injury of bladder
Torso	Severe	S37.3	Injury of urethra
Torso	Severe	S37.6	Injury of uterus
Torso	Severe	S37.7	Injury of multiple pelvic organs
Torso	Severe	S37.8	Injury of other pelvic organs
Torso	Severe	S37.9	Injury of unspecified pelvic organ
Torso	Severe	S38.0	Crushing injury of external genital organs
Torso	Severe	S38.1	Crushing of o. & u. parts of abdomen, lower back and pelvis
Torso	Severe	S38.2	Traumatic amputation of external genital organs
Torso	Severe	S38.3	Traum. amputation of o. & u. parts of abdomen, lower back and pelvis
Torso	Minor	S39.0	Injury of muscle and tendon of abdomen, lower back and pelvis
Torso	Severe	S39.6	Injury of intra-abdominal organ(s) with pelvic organ(s)
Torso	Unk Sev	S39.7	Other multiple injuries of abdomen, lower back and pelvis
Torso	Unk Sev	S39.8	Other specified injuries of abdomen, lower back and pelvis
Torso	Unk Sev	S39.9	Unspecified injury of abdomen, lower back and pelvis
Arm	Minor	S40.0	Contusion of shoulder and upper arm
Arm	Minor	S40.7	Multiple superficial injuries of shoulder and upper arm
Arm	Minor	S40.9	Superficial injury of shoulder and upper arm, unspecified
Arm	Minor	S41.0	Open wound of shoulder
Arm	Minor	S41.1	Open wound of upper arm
Arm	Minor	S41.8	Open wound of other and unspecified parts of shoulder girdle
Torso	Severe	S42.0	Fracture of clavicle
Torso	Severe	S42.1	Fracture of scapula
Arm	Minor	S42.2	Fracture of upper end of humerus
Arm	Minor	S42.3	Fracture of shaft of humerus
Arm	Minor	S42.4	Fracture of lower end of humerus
Arm	Severe	S42.7	Multiple fractures of clavicle, scapula and humerus
Arm	Unk Sev	S42.8	Fracture of other parts of shoulder and upper arm
Torso	Severe	S42.9	Fracture of shoulder girdle, part unspecified
Arm	Minor	S43.0	Dislocation of shoulder joint
Arm	Minor	S43.1	Dislocation of acromioclavicular joint
Torso	Severe	S43.2	Dislocation of sternoclavicular joint
Arm	Severe	S45.0	Injury of axillary artery
Arm	Severe	S45.1	Injury of brachial artery
Arm	Severe	S45.2	Injury of axillary or brachial vein
Arm	Severe	S45.9	Injury of unspecified blood vessel at shoulder and upper arm level
Arm	Severe	S47	Crushing injury of shoulder and upper arm
Arm	Severe	S48.0	Traumatic amputation at shoulder joint
Arm	Severe	S48.9	Traumatic amputation of shoulder and upper arm, level unspecified
Arm	Unk Sev	S49.7	Multiple injuries of shoulder and upper arm
Arm	Unk Sev	S49.8	Other specified injuries of shoulder and upper arm
Arm	Unk Sev	S49.9	Unspecified injury of shoulder and upper arm
Arm	Minor	S50.7	Multiple superficial injuries of forearm
Arm	Minor	S50.9	Superficial injury of forearm, unspecified
Arm	Minor	S51.0	Open wound of elbow
Arm	Minor	S51.9	Open wound of forearm, part unspecified
Arm	Minor	S52.0	Fracture of upper end of ulna
Arm	Minor	S52.1	Fracture of upper end of radius
Arm	Minor	S52.2	Fracture of shaft of ulna
Arm	Severe	S52.4	Fracture of shafts of both ulna and radius
Arm	Minor	S52.5	Fracture of lower end of radius
Arm	Severe	S52.7	Multiple fractures of forearm
Arm	Minor	S52.8	Fracture of other parts of forearm
Arm	Minor	S52.9	Fracture of forearm, part unspecified
Arm	Minor	S53.1	Dislocation of elbow, unspecified
Arm	Severe	S55.1	Injury of radial artery at forearm level
Arm	Severe	S58.9	Traumatic amputation of forearm, level unspecified
Arm	Unk Sev	S59.7	Multiple injuries of forearm
Arm	Unk Sev	S59.9	Unspecified injury of forearm
Arm	Minor	S60.8	Other superficial injuries of wrist and hand
Arm	Minor	S60.9	Superficial injury of wrist and hand, unspecified
Arm	Minor	S61.9	Open wound of wrist and hand part, unspecified
Arm	Minor	S62.1	Fracture of other carpal bone(s)
Arm	Minor	S62.3	Fracture of other metacarpal bone
Arm	Minor	S62.5	Fracture of thumb
Arm	Minor	S62.6	Fracture of other finger
Arm	Minor	S62.8	Fracture of other and unspecified parts of wrist and hand

REGION	SEVERITY	CODE	INJURY
Arm	Minor	S63.0	Dislocation of wrist
Arm	Minor	S68.1	Traumatic amputation of other single finger
Arm	Severe	S68.4	Traumatic amputation of hand at wrist level
Arm	Severe	S68.9	Traumatic amputation of wrist and hand, level unspecified
Arm	Unk Sev	S69.7	Multiple injuries of wrist and hand
Arm	Unk Sev	S69.9	Unspecified injury of wrist and hand
Torso	Minor	S70.0	Contusion of hip
Leg	Minor	S70.1	Contusion of thigh
Leg	Minor	S70.7	Multiple superficial injuries of hip and thigh
Leg	Minor	S70.8	Other superficial injuries of hip and thigh
Leg	Minor	S70.9	Superficial injury of hip and thigh, unspecified
Torso	Minor	S71.0	Open wound of hip
Leg	Minor	S71.1	Open wound of thigh
Torso	Severe	S72.0	Fracture of neck of femur/Fracture of hip
Leg	Severe	S72.1	Trochanteric fracture of femur
Leg	Severe	S72.2	Subtrochanteric fracture of femur
Leg	Severe	S72.3	Fracture of shaft of femur
Leg	Severe	S72.4	Fracture of lower end of femur
Leg	Severe	S72.7	Multiple fractures of femur
Leg	Severe	S72.8	Fractures of other parts of femur
Leg	Severe	S72.9	Fracture of femur, part unspecified
Torso	Minor	S73.0	Dislocation of hip
Leg	Severe	S75.0	Injury of femoral artery
Leg	Severe	S75.1	Injury of femoral vein at hip and thigh level
Leg	Severe	S75.9	Injury of unspecified blood vessel at hip and thigh level
Leg	Minor	S76.0	Injury of muscle and tendon of hip
Leg	Severe	S77.1	Crushing injury of thigh
Leg	Severe	S78.1	Traumatic amputation of thigh at level between hip and knee
Leg	Severe	S78.9	Traumatic amputation of hip and thigh, level unspecified
Leg	Unk Sev	S79.7	Multiple injuries of hip and thigh
Leg	Unk Sev	S79.8	Other specified injuries of hip and thigh
Leg	Unk Sev	S79.9	Unspecified injury of hip and thigh
Leg	Minor	S80.1	Contusion of other and unspecified parts of lower leg
Leg	Minor	S80.7	Multiple superficial injuries of lower leg
Leg	Minor	S80.8	Other superficial injuries of lower leg
Leg	Minor	S80.9	Superficial injury of lower leg, unspecified
Leg	Minor	S81.0	Open wound of knee
Leg	Minor	S81.8	Open wound of other parts of lower leg
Leg	Minor	S81.9	Open wound of lower leg, part unspecified
Leg	Severe	S82.0	Fracture of patella
Leg	Minor	S82.1	Fracture of upper end of tibia
Leg	Minor	S82.2	Fracture of shaft of tibia
Leg	Minor	S82.3	Fracture of lower end of tibia
Leg	Minor	S82.4	Fracture of fibula alone
Leg	Severe	S82.7	Multiple fractures of lower leg
Leg	Minor	S82.8	Fractures of other parts of lower leg
Leg	Minor	S82.9	Fracture of lower leg, part unspecified
Leg	Minor	S83.0	Dislocation of patella
Leg	Minor	S83.1	Dislocation of knee
Leg	Minor	S83.2	Tear of meniscus of knee, current
Leg	Severe	S83.7	Injury to multiple structures of knee
Leg	Severe	S84.7	Injury of multiple nerves at lower leg level
Leg	Severe	S85.0	Injury of popliteal artery
Leg	Severe	S85.1	Injury of (anterior)(posterior) tibial artery
Leg	Severe	S85.2	Injury of peroneal artery
Leg	Severe	S85.5	Injury of popliteal vein
Leg	Severe	S85.9	Injury of unspecified blood vessel at lower leg level
Leg	Minor	S86.0	Injury of Achilles tendon
Leg	Severe	S87.8	Crushing injury of other and unspecified parts of lower leg
Leg	Severe	S88.0	Traumatic amputation of leg at knee level
Leg	Severe	S88.1	Traumatic amputation of leg at level between knee and ankle
Leg	Severe	S88.9	Traumatic amputation of lower leg, level unspecified
Leg	Unk Sev	S89.7	Multiple injuries of lower leg
Leg	Unk Sev	S89.9	Unspecified injury of lower leg
Leg	Minor	S90.0	Contusion of ankle
Leg	Minor	S90.9	Superficial injury of ankle and foot, unspecified
Leg	Minor	S91.0	Open wound of ankle
Leg	Minor	S91.3	Open wound of other parts of foot

REGION	SEVERITY	CODE	INJURY
Leg	Minor	S92.0	Fracture of calcaneus of heel
Leg	Minor	S92.1	Fracture of talus of foot
Leg	Minor	S92.2	Fracture of other tarsal bone(s)
Leg	Minor	S92.3	Fracture of metatarsal bone
Leg	Minor	S92.5	Fracture of non-big toe
Leg	Minor	S92.7	Multiple fractures of foot
Leg	Minor	S92.9	Fracture of foot, unspecified
Leg	Minor	S93.0	Dislocation of ankle joint
Leg	Minor	S93.3	Dislocation of other and unspecified parts of foot
Leg	Minor	S96.1	Injury of long extensor muscle of toe at ankle and foot level
Leg	Minor	S96.8	Injury of other muscles and tendons at ankle and foot level
Leg	Severe	S97.0	Crushing injury of ankle
Leg	Severe	S98.0	Traumatic amputation of foot at ankle level
Leg	Minor	S98.2	Traumatic amputation of two or more toes
Leg	Severe	S98.4	Traumatic amputation of foot, level unspecified
Leg	Unk Sev	S99.9	Unspecified injury of ankle and foot
Arm	Minor	T00.2	Superficial injuries involving multiple regions of upper limb(s)
Leg	Minor	T00.3	Superficial injuries involving multiple regions of lower limb(s)
Multi	Minor	T00.8	Superficial injuries involving other combinations of body regions
Unk.	Minor	T00.9	Multiple superficial injuries, unspecified
Arm	Minor	T01.2	Open wounds involving multiple regions of upper limb(s)
Leg	Minor	T01.3	Open wounds involving multiple regions of lower limb(s)
Unk.	Minor	T01.9	Multiple open wounds, unspecified
Torso	Severe	T02.1	Fractures involving thorax with lower back and pelvis
Arm	Severe	T02.2	Fractures involving multiple regions of one upper limb
Leg	Severe	T02.3	Fractures involving multiple regions of one lower limb
Arm	Severe	T02.4	Fractures involving multiple regions of both upper limbs
Leg	Severe	T02.5	Fractures involving multiple regions of both lower limbs
Multi	Severe	T02.8	Fractures involving other combinations of body regions
Multi	Severe	T02.9	Multiple fractures, unspecified
Multi	Unk Sev	T03.8	Disloc's, sprains and strains involving combinations of body regions
Unk.	Unk Sev	T03.9	Multiple dislocations, sprains and strains, unspecified
Torso	Severe	T04.1	Crushing injuries involving thorax with abdomen, lower back & pelvis
Arm	Severe	T04.2	Crushing injuries involving multiple regions of upper limb(s)
Leg	Severe	T04.3	Crushing injuries involving multiple regions of lower limb(s)
Multi	Severe	T04.8	Crushing injuries involving other combinations of body regions
Multi	Severe	T04.9	Multiple crushing injuries, unspecified
Leg	Severe	T05.5	Traumatic amputation of both legs [any level]
Multi	Severe	T05.8	Traumatic amputations involving other combinations of body regions
Multi	Severe	T05.9	Multiple traumatic amputations, unspecified
Multi	Unk Sev	T06.2	Injuries of nerves involving multiple body regions
Multi	Severe	T06.3	Injuries of blood vessels involving multiple body regions
Multi	Unk Sev	T06.4	Injuries of muscles and tendons involving multiple body regions
Torso	Severe	T06.5	Injuries of intrathoracic, intra-abdominal & pelvic organs
Unk.	Unk Sev	T07	Unspecified multiple injuries
Unk.	Severe	T08	Fracture of spine, level unspecified
Torso	Minor	T09.0	Superficial injury of trunk, level unspecified
Torso	Minor	T09.1	Open wound of trunk, level unspecified
Torso	Unk Sev	T09.2	Dislocation, sprain and strain of u. joint and ligament of trunk
Unk.	Severe	T09.3	Injury of spinal cord, level unspecified
Torso	Unk Sev	T09.4	Injury of unspecified nerve, spinal nerve root and plexus of trunk
Torso	Minor	T09.5	Injury of unspecified muscle and tendon of trunk
Torso	Severe	T09.6	Traumatic amputation of trunk, level unspecified
Torso	Unk Sev	T09.8	Other specified injuries of trunk, level unspecified
Torso	Unk Sev	T09.9	Unspecified injury of trunk, level unspecified
Arm	Unk Sev	T10	Fracture of upper limb, level unspecified
Arm	Minor	T11.0	Superficial injury of upper limb, level unspecified
Arm	Minor	T11.1	Open wound of upper limb, level unspecified
Arm	Minor	T11.2	Disloc, sprain, strain of u. joint/ligament of upper limb, level unk.
Arm	Severe	T11.4	Injury of unspecified blood vessel of upper limb, level unspecified
Arm	Minor	T11.5	Injury of u. muscle and tendon of upper limb, level unspecified
Arm	Severe	T11.6	Traumatic amputation of upper limb, level unspecified
Arm	Unk Sev	T11.8	Other specified injuries of upper limb, level unspecified
Arm	Unk Sev	T11.9	Unspecified injury of upper limb, level unspecified
Leg	Unk Sev	T12	Fracture of lower limb, level unspecified
Leg	Minor	T13.0	Superficial injury of lower limb, level unspecified
Leg	Minor	T13.1	Open wound of lower limb, level unspecified
Leg	Minor	T13.2	Disloc, sprain, strain of u. joint/ligament of lower limb, level unk.

REGION	SEVERITY	CODE	INJURY
Leg	Severe	T13.4	Injury of unspecified blood vessel of lower limb, level unspecified
Leg	Minor	T13.5	Injury of u. muscle and tendon of lower limb, level unspecified
Leg	Severe	T13.6	Traumatic amputation of lower limb, level unspecified
Leg	Unk Sev	T13.8	Other specified injuries of lower limb, level unspecified
Leg	Unk Sev	T13.9	Unspecified injury of lower limb, level unspecified
Unk.	Minor	T14.0	Superficial injury of unspecified body region
Unk.	Minor	T14.1	Open wound of unspecified body region
Unk.	Unk Sev	T14.2	Fracture of unspecified body region
Unk.	Unk Sev	T14.3	Dislocation, sprain and strain of unspecified body region
Unk.	Unk Sev	T14.4	Injury of nerve(s) of unspecified body region
Unk.	Unk Sev	T14.5	Injury of blood vessel(s) of unspecified body region
Unk.	Minor	T14.6	Injury of muscles and tendons of unspecified body region
Unk.	Severe	T14.7	Crushing injury and traumatic amputation of unspecified body region
Unk.	Unk Sev	T14.8	Other injuries of unspecified body region
Unk.	Unk Sev	T14.9	Injury, unspecified
Head	Minor	T15.1	Foreign body in conjunctival sac of eye
Head	Minor	T15.9	Foreign body on external eye, part unspecified
Head	Minor	T17.1	Foreign body in nostril
Neck	Unk Sev	T17.3	Foreign body in larynx
Neck	Unk Sev	T17.4	Foreign body in trachea
Torso	Unk Sev	T17.5	Foreign body in bronchus
Unk.	Unk Sev	T17.8	Foreign body in other and multiple parts of respiratory tract
Unk.	Unk Sev	T17.9	Foreign body in respiratory tract, part unspecified
Unk.	Unk Sev	T18.9	Foreign body in alimentary tract, part unspecified
Burn	Any	T20-T31	ALL BURNS
N/A	Unk Sev	T35.5	Unspecified frostbite of lower limb
N/A	Unk Sev	T36-T50	Poisoning by medicines or recreational drugs
Other	Unk Sev	T52.0	Toxic effect of petroleum products
N/A	Unk Sev	T56.8	Toxic effect of other metals
N/A	Unk Sev	T57.1	Toxic effect of phosphorus and its compounds
Other	Unk Sev	T58	Toxic effect of carbon monoxide
N/A	Unk Sev	T59.1	Toxic effect of sulfur dioxide
N/A	Unk Sev	T59.7	Toxic effect of carbon dioxide
Other	Unk Sev	T59.8	Toxic effect of other specified gases, fumes and vapors
Other	Unk Sev	T59.9	Toxic effect of gases, fumes and vapors, unspecified
N/A	Unk Sev	T63.4	Toxic effect of insect bites
Other	Unk Sev	T65.9	Toxic effect of unspecified substance
N/A	Unk Sev	T67.0	Heatstroke and sunstroke
N/A	Unk Sev	T67.9	Effect of heat and light, unspecified
N/A	Unk Sev	T68	Hypothermia
N/A	Unk Sev	T69.8	Other specified effects of reduced temperature
N/A	Unk Sev	T69.9	Effect of reduced temperature, unspecified
Other	Unk Sev	T70.9	Effect of air pressure and water pressure, unspecified
Other	Severe	T71	Asphyxiation
N/A	Unk Sev	T73.8	Other effects of deprivation (exposure, overexertion)
Other	Severe	T75.1	Drowning
N/A	Unk Sev	T75.4	Effects of electric current
N/A	Unk Sev	T75.8	Other specified effects of external causes
N/A	Unk Sev	T78.2	Anaphylactic shock
N/A	Unk Sev	T78.4	Allergy, unspecified
N/A	Unk Sev	T79.0	Air embolism (traumatic)
N/A	Unk Sev	T79.1	Fat embolism (traumatic)
N/A	Unk Sev	T79.2	Traumatic secondary and recurrent hemorrhage
N/A	Unk Sev	T79.3	Post-traumatic wound infection
N/A	Unk Sev	T79.4	Traumatic shock
N/A	Unk Sev	T79.5	Traumatic anuria
N/A	Unk Sev	T79.6	Traumatic ischemia of muscle
N/A	Unk Sev	T79.7	Traumatic subcutaneous emphysema
N/A	Unk Sev	T79-T88	Complications of trauma or procedures
N/A	Unk Sev	T90-T98	Sequelae of injuries, toxic effects, complications

APPENDIX C

MAKE-MODEL GROUPS FOR EVALUATING FMVSS No. 201 MODEL-YEAR RANGES INCLUDED IN THE FARS-MCOD ANALYSES

2300 Jeep Cherokee
Includes make-models: 2300-2303 (based on the 5-digit make-model code, MM2, generated by the VIN analysis programs developed by NHTSA staff for use in evaluations of FMVSS and available to the public at www.nhtsa.gov/fuel-economy)
FMVSS No. 201 certification: never
Produced: 1996-2001
Side air bags: never
Frontal air bags: driver 1996, dual 1997-2001
Pretensioners: never
Major redesigns: none
Decision: generally exclude because cannot combine with Jeep Liberty (major redesign + pretensioners added). However, for unrestrained occupants, maybe combine with Jeep Liberty if the latter 201-certified as early as 2002; compare 1999-2001 Cherokee to 2002-2004 Liberty.

2312 Jeep Grand Cherokee
Includes make-models: 2312-2313
FMVSS No. 201 certification: mid-year 1999
Produced: always
Side air bags: a small proportion of VIN-identified head curtains only in 2002-2004
Frontal air bags: always
Pretensioners: 2004 only
Major redesigns: none
Decision: use 1996-1998 for "before" and 2000-2002 for "after," excluding vehicles with head curtains

2321 Jeep Wrangler
Includes make-models: 2321
FMVSS No. 201 certification: 2003
Produced: always
Side air bags: never
Frontal air bags: always
Pretensioners: never
Major redesigns: LWB model introduced in 2004 while old model continued (no 2004 LWB on FARS)
Decision: use 2000-2002 for "before" and 2003-2005 for "after" but exclude LWB models (CG = 6308; CG is the 5-digit vehicle-group code generated by the VIN analysis programs available to the public at www.nhtsa.gov/fuel-economy)

2342 Jeep Liberty
Includes make-models: 2342-2343
FMVSS No. 201 certification: began 9/1/2001 (production began 8/1/2001, probably reasonable to assume all Jeep Liberty met 201, because unlikely to have modified it after a month, non-certification was probably a regulatory convenience)
Produced: 2002-2004
Side air bags: a small proportion of VIN-identified head curtains only in 2002-2004
Frontal air bags: always
Pretensioners: always
Major redesigns: none
Decision: generally exclude because cannot combine with Jeep Cherokee (major redesign + pretensioners added); however, for unrestrained occupants, compare 1999-2001 Cherokee to 2002-2004 Liberty, excluding vehicles with head curtains.

6041 Chrysler Concorde
Includes make-models: 6041 (Concorde), 6042 (LHS)
FMVSS No. 201 certification: mid-year 1999
Produced: always
Side air bags: a small proportion of VIN-identified combination bags in 2001-2004
Frontal air bags: always
Pretensioners: never
Major redesigns: none
Decision: use 1997-1998 as "before" and 2000-2002 as "after," excluding vehicles with combination bags

6043 Chrysler Sebring convertible
Includes make-models: 6043 (Chrysler Sebring) convertible only
FMVSS No. 201 certification: 2002
Produced: always
Side air bags: never
Frontal air bags: always
Pretensioners: 2001-2004
Major redesigns: none
Decision: use 2001 as "before" and 2002 as "after"; for unrestrained occupants, use 2000-2001 as "before" and 2002-2004 as "after."

6052 Chrysler PT Cruiser SUV
Includes make-models: 6052
FMVSS No. 201 certification: 2003
Produced: 2001-2005
Side air bags: substantial proportions of VIN-identified combination bags in 2001-2004
Frontal air bags: always
Pretensioners: always
Major redesigns: none
Decision: use 2001-2002 as "before" and 2003-2005 as "after," excluding vehicles with combination bags; also exclude convertible (CG=6034)

7013 Dodge Viper
Includes make-models: 7013
FMVSS No. 201 certification: 2003
Produced: always
Side air bags: never
Frontal air bags: 1997-2005
Pretensioners: 2003-2005
Major redesigns: 2003
Decision: generally exclude because pretensioners added in 2003; however, for unrestrained occupants, use 2000-2002 for "before" and 2003-2005 for "after"

7020 Dodge Neon
Includes make-models: 7020 (Dodge Neon), 9020 (Plymouth Neon)
FMVSS No. 201 certification: some cars in 2000 and 2001, all in 2002
Produced: always
Side air bags: a small proportion of VIN-identified combination bags in 2001-2004
Frontal air bags: always
Pretensioners: never
Major redesigns: 2000
Decision: use 1998-1999 as "before" and 2002-2004 as "after," excluding vehicles with combination bags

7041 Dodge Intrepid
Includes make-models: 6051 (Chrysler 300), 7041 (Dodge Intrepid)
FMVSS No. 201 certification: some cars in 1999 and 2000, all in 2001
Produced: always
Side air bags: a small proportion of VIN-identified combination bags in 2001-2004
Frontal air bags: always
Pretensioners: never
Major redesigns: none
Decision: use 1997-1998 as "before" and 2001-2003 as "after," excluding vehicles with combination bags

7042 Dodge Avenger 2-door coupe
Includes make-models: 6043 (Chrysler Sebring 2-door coupe only), 7042 (Dodge Avenger), 7043 (Dodge Stratus 2-door coupe only)
FMVSS No. 201 certification: 2002
Produced: always
Side air bags: never
Frontal air bags: always
Pretensioners: never
Major redesigns: none
Decision: use 1999-2001 as "before" and 2002-2004 as "after."

7043 Dodge Stratus 4-door sedan
Includes make-models: 6043 (Chrysler Sebring 4-door sedan only), 6044 (Chrysler Cirrus), 7043 (Dodge Stratus 4-door sedan only), 9038 (Plymouth Breeze)
FMVSS No. 201 certification: 2001
Produced: always
Side air bags: a small proportion of VIN-identified head curtains only in 2002-2004
Frontal air bags: always
Pretensioners: 2001-2004
Major redesigns: none
Decision: generally exclude because pretensioners added in 2001; however, for unrestrained occupants, use 1998-2000 for "before" and 2001-2003 for "after," excluding vehicles with head curtains

7200 Dodge Dakota conventional or club cab
Includes make-models: 7200-7203
FMVSS No. 201 certification: 2000
Produced: always
Side air bags: never
Frontal air bags: driver only in 1996, dual without switches in 1997, dual with on-off switch in 1998-2004
Pretensioners: never
Major redesigns: none
Decision: use 1998-1999 for "before" and 2000-2002 for "after"

7204 Dodge Dakota quad cab
Includes make-models: 7204-7205
FMVSS No. 201 certification: 2000
Produced: 2000-2004
Side air bags: never
Frontal air bags: always
Pretensioners: 2002-2004
Major redesigns: none
Decision: exclude because no pre-FMVSS No. 201 trucks

7210 Dodge Ram 1500 pickup
Includes make-models: 7210-7215, 7510-7515
FMVSS No. 201 certification: 2002
Produced: always
Side air bags: a small proportion of VIN-identified head curtains only in 2002-2004
Frontal air bags: driver only in 1996-1997, dual with on-off switch in 1998-2004 (except quad cabs in 2002-2004), dual in 2002-2004 (quad cabs only)
Pretensioners: 2002-2004
Major redesigns: 2002
Decision: generally exclude because pretensioners added in 2002; however, for unrestrained occupants, use 2000-2001 for "before" and 2002-2004 for "after," excluding vehicles with head curtains

7220 Dodge Ram 2500/3500 pickup
Includes make-models: 7220-7239, 7520-7539, excluding vehicles over 10,000 GVWR
FMVSS No. 201 certification: 2003
Produced: always
Side air bags: a small proportion of VIN-identified head curtains only in 2003-2005
Frontal air bags: driver only in 1996-1997, dual with on-off switch in 1998-2005 (except quad cabs in 2003-2004), dual in 2003-2005 (quad cabs only)
Pretensioners: 2003-2005
Major redesigns: 2003
Decision: generally exclude because pretensioners added in 2003 however, for unrestrained occupants, use 2000-2002 for "before" and 2003-2005 for "after," excluding vehicles with head curtains

7312 Dodge Durango
Includes make-models: 7312, 7313
FMVSS No. 201 certification: 2001
Produced: 1998-2004
Side air bags: a small proportion of VIN-identified head curtains only in 2002-2004
Frontal air bags: always
Pretensioners: 2001-2004
Major redesigns: 2004
Decision: generally exclude because pretensioners added in 2002; however, for unrestrained occupants, use 1999-2000 for "before" and 2001-2003 for "after," excluding vehicles with head curtains

7402 Dodge Caravan
Includes make-models: 6400-6409, 7400-7409, 7600-7609, 9400-9409 (Caravan, Grand Caravan, Voyager, Grand Voyager, Town & Country)
FMVSS No. 201 certification: 2002 (however, even though Chrysler confirms that not certified in 2001, I will skip this year because it is likely that vehicles were modified upon redesign in 2001 and simply not certified until 2002)
Produced: always
Side air bags: a moderate proportion of VIN-identified combination bags in 2001-2004
Frontal air bags: always
Pretensioners: 2001-2004, upon redesign
Major redesigns: 2001
Decision: generally exclude because pretensioners added in 2001; however, for unrestrained occupants, use 1999-2000 for "before" and 2002-2004 for "after," excluding vehicles with combination bags

7410 Dodge Ram Van
Includes make-models: 7410-7439, 7610-7639
FMVSS No. 201 certification: 2003
Produced: 1996-2003
Side air bags: never
Frontal air bags: driver air bags 1996-1997, dual air bags 1998-2003
Pretensioners: 2001-2003
Major redesigns: 1998
Decision: exclude because few FMVSS No. 201 vans

12003.1 Ford Mustang convertible
Includes make-models: 12003 with BOD2=1 (the body-type code for convertibles in NHTSA's VIN-analysis programs)
FMVSS No. 201 certification: 2001
Produced: always
Side air bags: never
Frontal air bags: always
Pretensioners: never
Major redesigns: none
Decision: use 1998-2000 for "before" and 2001-2003 for "after"

12003.2 Ford Mustang coupe
Includes make-models: 12003 with BOD2=2 (body-type code for 2-door coupe)
FMVSS No. 201 certification: 2003
Produced: always
Side air bags: combo bags 2005 if V4=H (4th character of VIN)
Frontal air bags: always
Pretensioners: 2005
Major redesigns: 2005
Decision: use 2001-2002 for "before" and 2003-2004 for 'after"; for unrestrained, use 2000-2002 for "before" and 2003-2005 for "after," excluding cars with combo bags

12004 Ford Thunderbird
Includes make-models: 12004
FMVSS No. 201 certification: 2002
Produced: 1996-1997 and 2002-2004
Side air bags: combination bags standard in 2002-2004
Frontal air bags: always
Pretensioners: 2002-2004
Major redesigns: 2002
Decision: exclude because no recent pre-FMVSS No. 201 cars

12013 Ford Escort
Includes make-models: 12013
FMVSS No. 201 certification: 2000
Produced: 1996-2003
Side air bags: never
Frontal air bags: always
Pretensioners: never
Major redesigns: none
Decision: use 1998-1999 for "before" and 2000-2002 for "after"

12017 Ford Taurus
Includes make-models: 12017 (Ford Taurus), 14017 (Mercury Sable)
FMVSS No. 201 certification: 2000
Produced: always
Side air bags: a small proportion of VIN-identified combination bags in 2001-2004
Frontal air bags: always
Pretensioners: 2000-2004
Major redesigns: none
Decision: generally exclude because pretensioners added in 2000; however, for unrestrained occupants, use 1997-1999 for "before" and 2000-2002 for "after," excluding vehicles with combination bags

12037 Ford Focus
Includes make-models: 12037
FMVSS No. 201 certification: 2000
Produced: 2000-2004
Side air bags: a small proportion of VIN-identified combination bags in 2000-2004
Frontal air bags: always
Pretensioners: 2000-2004
Major redesigns: none
Decision: exclude because no pre-FMVSS No. 201 cars (Ford Contour/Mercury Mystique is not a good pre-FMVSS No. 201 comparison group because Focus was intended more as a replacement for the Escort than the Contour/Mystique)

12200 Ford Ranger
Includes make-models: 12200-12205 (Ford Ranger), 41200-41205 (Mazda B-Pickup)
FMVSS No. 201 certification: 2003
Produced: always
Side air bags: never
Frontal air bags: driver air bags (with optional passenger bag plus on-off switch) 1996-1997, dual air bags with on-off switch, 1998-2005
Pretensioners: 2001-2005
Major redesigns: 1998
Decision: use 2001-2002 for "before" and 2003-2004 for "after"; for unrestrained occupants use 2000-2002 for "before" and 2003-2005 for "after"

12210 Ford F-150 pickup
Includes make-models: 12210-12219
FMVSS No. 201 certification: 2002
Produced: always
Side air bags: never
Frontal air bags: driver air bags 1996, dual air bags with on-off switch (except crew cabs), 1997-2004, crew cabs have dual air bags, 2001-2004
Pretensioners: 2001-2004
Major redesigns: mid-1997, mid-2004
Decision: use 2001 for "before" and 2002-2003 for "after"; for unrestrained occupants, use 2000-2001 for "before" and 2002-2004 for "after" (pretensioners not an issue)

12220 Ford F-250/350 pickup
Includes make-models: 12220-12239 and 12520-12539 with GVWR under 10,000 pounds
FMVSS No. 201 certification: 2003
Produced: always
Side air bags: never
Frontal air bags: various 1996-1997, dual air bags with on-off switch (except crew cabs), 1998-2004, crew cabs have dual air bags, 2000-2004, all have dual CAC air bags 2005
Pretensioners: never according to cars.com
Major redesigns: 1999
Decision: use 2000-2002 for "before" and 2003-2005 for "after"

12300 Ford Explorer 2-door
Includes make-models: 12300, 12301
FMVSS No. 201 certification: 2002
Produced: 1996-2003
Side air bags: a small proportion of VIN-identified combination bags in 1999-2003
Frontal air bags: always
Pretensioners: 2003 only
Major redesigns: none
Decision: use 2001 for "before" and 2002 for "after," excluding vehicles with combination bags; for unrestrained occupants, use 2001 for "before" and 2002-2003 for "after" (pretensioners not an issue)

12302 Ford Explorer 4-door
Includes make-models: 12302, 12303, 12308 (Ford Explorer); 14302, 14303, 14308 (Mercury Mountaineer) [Lincoln Aviator excluded because combination bags are standard]
FMVSS No. 201 certification: 2002
Produced: always
Side air bags: small VIN-identified proportions of combination bags in 1999-2001 and head curtains 2002-2004
Frontal air bags: always
Pretensioners: 2002-2004
Major redesigns: 2002
Decision: generally exclude because pretensioners added in 2002; however, for unrestrained occupants, use 1999-2001 for "before" and 2002-2004 for "after," excluding vehicles with combination bags or head curtains

12306 Ford Explorer Sport-Trac
Includes make-models: 12306, 12307
FMVSS No. 201 certification: 2002
Produced: 2001-2004
Side air bags: a small proportion of VIN-identified head-curtains in 2003-2004
Frontal air bags: always
Pretensioners: 2003-2004
Major redesigns: none
Decision: use 2001 for "before" and 2002 for "after"; for unrestrained occupants, use 2001 for "before" and 2002-2004 for "after," excluding vehicles with head curtains (pretensioners not an issue)

12312 Ford Expedition
Includes make-models: 12312, 12313
FMVSS No. 201 certification: 2003
Produced: 1997-2005
Side air bags: small VIN-identified proportions of combination bags in 2000-2002 and head curtains 2003-2005
Frontal air bags: always
Pretensioners: 2001-2005
Major redesigns: none
Decision: use 2001-2002 for "before" and 2003-2004 for "after," excluding vehicles with combination bags or head curtains; for unrestrained use 2000-2002 for "before" and 2003-2005 for "after," excluding vehicles with combination bags or head curtains

12332 Ford Excursion
Includes make-models: 12332, 12333
FMVSS No. 201 certification: 2003
Produced: 2000-2005
Side air bags: never
Frontal air bags: always
Pretensioners: 2002-2005
Major redesigns: none
Decision: use 2002 for "before" and 2003-2004 for "after"; for unrestrained occupants, use 2000-2002 for "before" and 2003-2005 for "after" (pretensioners not an issue)

12342 Ford Escape
Includes make-models: 12342, 12343 (Ford Escape); 41342, 41343 (Mazda Tribute)
FMVSS No. 201 certification: 2001
Produced: 2001-2004
Side air bags: a small proportion of VIN-identified combination bags in 2001-2004
Frontal air bags: always
Pretensioners: 2001-2004
Major redesigns: none
Decision: exclude because no pre-FMVSS No. 201 SUVs

12402 Ford Windstar
Includes make-models: 12400, 12402
FMVSS No. 201 certification: 1999
Produced: 1996-2003
Side air bags: a small proportion of VIN-identified combination bags in 1999-2003
Frontal air bags: always
Pretensioners: 2000-2003
Major redesigns: 1999 [and replaced in 2004 by Ford Freestar with the same MM2 code]
Decision: use 1998 for "before" and 1999 for "after," excluding vehicles with combination bags; for unrestrained occupants, use 1996-1998 for "before" and 1999-2001 for "after," excluding vehicles with combination bags (pretensioners not an issue)

12410 Ford full-size vans
Includes make-models: 12410-12439, 12610-12639 with GVWR under 10,000 pounds
FMVSS No. 201 certification: 2002
Produced: always
Side air bags: never
Frontal air bags: driver air bags 1996, dual air bags 1997-2004
Pretensioners: 1998-2004
Major redesigns: none
Decision: use 1999-2001 for "before" and 2002-2004 for "after"

13001 Lincoln Town Car
Includes make-models: 13001
FMVSS No. 201 certification: 2001
Produced: always
Side air bags: combination bags standard in 1999-2004
Frontal air bags: always
Pretensioners: 2001-2004
Major redesigns: none
Decision: exclude because all FMVSS No. 201-certified cars are equipped with head-protection air bags

13005 Lincoln Continental
Includes make-models: 13005
FMVSS No. 201 certification: 2002
Produced: 1996-2002
Side air bags: combination bags standard in 1999-2002
Frontal air bags: always
Pretensioners: 2002 only
Major redesigns: none
Decision: exclude because all FMVSS No. 201-certified cars are equipped with head-protection air bags

13012 Lincoln LS
Includes make-models: 13012 (Lincoln LS), 39034 (Jaguar S-Type)
FMVSS No. 201 certification: 2000
Produced: 2000-2004
Side air bags: combination bags or head curtains plus torso bags standard in 2000-2004
Frontal air bags: always
Pretensioners: always
Major redesigns: none
Decision: exclude because no pre-FMVSS No. 201 cars

13312 Lincoln Navigator
Includes make-models: 13312, 13313
FMVSS No. 201 certification: 2003
Produced: 1998-2004
Side air bags: combination bags or head curtains standard in 2000-2004
Frontal air bags: always
Pretensioners: 2001-2004
Major redesigns: none
Decision: exclude because all FMVSS No. 201-certified SUVs are equipped with head-protection air bags

14016 Mercury Grand Marquis
Includes make-models: 12016 (Ford Crown Victoria), 14016 (Mercury Grand Marquis)
Excludes make-model: 14039 (Mercury Marauder) because combination bags are standard
FMVSS No. 201 certification: 2001
Produced: always
Side air bags: a small proportion of VIN-identified combination bags in 2003-2004
Frontal air bags: always
Pretensioners: 2001-2004
Major redesigns: none
Decision: generally exclude because pretensioners added in 2001; however, for unrestrained occupants, use 1998-2000 for "before" and 2001-2003 for "after," excluding vehicles with combination bags

14038 Mercury Cougar
Includes make-models: 14038
FMVSS No. 201 certification: 2002
Produced: 1999-2002
Side air bags: a small proportion of VIN-identified combination bags in 1999-2002
Frontal air bags: always
Pretensioners: 2002 only
Major redesigns: none
Decision: generally exclude because pretensioners added in 2002; however, for unrestrained occupants, use 2001 for "before" and 2002 for "after," excluding vehicles with combination bags

14452 Mercury Villager van
Includes make-models: 14450, 14452 (Mercury Villager); 35452 (Nissan Quest)
FMVSS No. 201 certification: 1999
Produced: 1996-2002, 2004
Side air bags: none in 1996-2002
Frontal air bags: always
Pretensioners: 2001-2002, 2004
Major redesigns: 1999, 2004
Decision: use 2001 for "before" and 2002 for "after"

18002 Buick LeSabre
Includes make-models: 18002 (Buick LeSabre), 22002 (Pontiac Bonneville)
FMVSS No. 201 certification: 2000
Produced: always
Side air bags: torso bags standard in 2000-2002; in 2003-2004 they became VIN-identified options on LeSabre but continued standard on Bonneville
Frontal air bags: always
Pretensioners: never
Major redesigns: 2000
Decision: exclude because FMVSS No. 201-certification coincided with standard torso bags

18003 Buick Park Avenue
Includes make-models: 18003
FMVSS No. 201 certification: 2003
Produced: always
Side air bags: torso bags standard in 2000-2005
Frontal air bags: always
Pretensioners: never
Major redesigns: 1997
Decision: use 2000-2002 for "before" and 2003-2005 for "after"

18017 Buick Century
Includes make-models: 18017 (Century), 18020 (Regal)
FMVSS No. 201 certification: 2003
Produced: Century -2005, Regal -2004
Side air bags: a moderate proportion of VIN-identified combo bags for drivers in 2000-2005
Frontal air bags: driver, 1996; dual 1997-2005
Pretensioners: never
Major redesigns: 1997
Decision: use 2000-2002 for "before" and 2003-2005 for "after," excluding cars with combination bags

18356 Buick Rendezvous
Includes make-models: 18356, 18357
FMVSS No. 201 certification: 2002
Produced: 2002-2005
Side air bags: combination bags for drivers and torso bags for RF standard or optional
Frontal air bags: always
Pretensioners: 2004 only
Major redesigns: none
Decision: exclude because no pre-FMVSS No. 201 SUVs

19003 Cadillac DeVille
Includes make-models: 19003
FMVSS No. 201 certification: 2000
Produced: always
Side air bags: torso bags standard in 1997-2004
Frontal air bags: always
Pretensioners: none thru 1999; maybe in 2000-2004 or maybe in 2000 only
Major redesigns: 2000
Decision: generally exclude because pretensioners may have been added in 2000; however, for unrestrained occupants, use 1998-1999 for "before" and 2000-2002 for "after"

19014 Cadillac Seville
Includes make-models: 19013
FMVSS No. 201 certification: 2002
Produced: always
Side air bags: torso bags standard in 1998-2000, combination bags for drivers and torso bags for RF standard in 2001-2004
Frontal air bags: always
Pretensioners: 1998-2004
Major redesigns: 1998
Decision: exclude because all FMVSS No. 201-certified cars are equipped with head-protection air bags [for drivers]

19312 Cadillac Escalade
Includes make-models: 19312, 19313 (Escalade); 19323 (Escalade ESV); 19343 (Escalade EXT)
FMVSS No. 201 certification: 2002
Produced: 1999-2000, 2002-2004
Side air bags: torso bags standard in 2002-2004
Frontal air bags: always
Pretensioners: never
Major redesigns: 2002
Decision: exclude because FMVSS No. 201-certification coincided with standard torso bags

20002 Chevrolet Impala
Includes make-models: 20002
FMVSS No. 201 certification: 2000
Produced: 2000-2004
Side air bags: a large proportion of VIN-identified combination bags for drivers in 2000-2004
Frontal air bags: always
Pretensioners: never
Major redesigns: none
Decision: exclude because no pre-FMVSS No. 201 cars

20004 Chevrolet Corvette
Includes make-models: 20004
FMVSS No. 201 certification: 2003
Produced: always
Side air bags: 2005 combo if V7=4
Frontal air bags: always
Pretensioners: 2005
Major redesigns: 2005
Decision: use 2001-2002 for "before" and 2003-2004 for "after"; for unrestrained use 2000-2002 for "before" and 2003-2005 for "after" and exclude combo bags

20016 Chevrolet Cavalier
Includes make-models: 20016 (Chevrolet Cavalier), 22016 (Pontiac Sunfire)
FMVSS No. 201 certification: 2003
Produced: always
Side air bags: a small proportion of VIN-identified torso bags in 2003-2005
Frontal air bags: always
Pretensioners: never
Major redesigns: none
Decision: use 2000-2002 for "before" and 2003-2005 for "after," excluding cars with torso bags

20036 Chevrolet Monte Carlo
Includes make-models: 20036
FMVSS No. 201 certification: 2000
Produced: always
Side air bags: a large proportion of VIN-identified combination bags for drivers in 2001-2002 and small
    proportions in 2003-2004
Frontal air bags: always
Pretensioners: never
Major redesigns: 2000
Decision: use 1998-1999 for "before" and 2000-2002 for "after," excluding cars with combination bags

20037 Chevrolet Malibu
Includes make-models: 20037 from CG 18068 only (Malibu in 1997-2003 and Classic in 2004-2005)
Excludes make-models: 20037 from CG 18078, 18079
FMVSS No. 201 certification: 2003
Produced: 1997-2005
Side air bags: never
Frontal air bags: always
Pretensioners: never
Major redesigns: none
Decision: use 2000-2002 for "before" and 2003-2005 for "after"

20200 GM S/T Pickup with conventional- or maxi-cab
Includes make-models: 20200-20203 (Chevrolet S/T) and 23200-23203 (GMC Sonoma) from CG 18205,
    18206 only
Excludes make-models: 20200-20203 (Colorado) and 23200-23203 (Canyon) from CG 18221, 18222
FMVSS No. 201 certification: 2003
Produced: 1996-2003
Side air bags: never
Frontal air bags: driver air bags in 1996-1997, dual air bags with on-off switches in 1998-2003
Pretensioners: never
Major redesigns: none
Decision: use 2002 for "before" and 2003 for "after"

20204 GM S/T Pickup with crew cab
Includes make-models: 20204-20205 (Chevrolet S/T) and 23204-23205 (GMC Sonoma) from CG 18206 only
Excludes make-models: 20204-20205 (Colorado) and 23204-23205 (Canyon) from CG 18222
FMVSS No. 201 certification: 2003
Produced: 2001-2004
Side air bags: never
Frontal air bags: always
Pretensioners: never
Major redesigns: none
Decision: use 2001-2002 for "before" and 2003-2004 for "after"

20210 GM 1500 pickup with conventional- or maxi-cab
Includes make-models: 20210-20213, 20510-20513 (Chevrolet C/K and Silverado); 23210-23213, 23510-23513 (GMC Sierra)
Excludes make-models: 1500 series crew cabs, because no pre-FMVSS No. 201 trucks; 2500 series trucks, because unknown if they had frontal air bags before Silverado redesign; 3500 series trucks, because always over 10,000 pounds GVWR after Silverado redesign
FMVSS No. 201 certification: CG 18213, 18215 (Silverado) are certified; CG 18207, 18210 (C/K) are not certified
Produced: always
Side air bags: never
Frontal air bags: driver air bags in 1996, dual air bags with on-off switches in 1997-2003
Pretensioners: never
Major redesigns: circa 1999 (from C/K to Silverado)
Decision: use 1997-1999 cases from CG 18207, 18210 for "before" and 1999-2000 cases from CG 18213, 18215 for "after"

20300 Chevrolet Blazer
Includes make-models: 20300, 20301 (2-door) from CG 18301; 20302, 20303 (4-door) from CG 18305 only
Excludes make-models: CG 18314 (Trailblazer)
FMVSS No. 201 certification: 2003
Produced: always
Side air bags: never
Frontal air bags: driver in 1996-1997; dual in 1998-2005
Pretensioners: never
Major redesigns: none
Decision: use 2001-2002 for "before" and 2003-2005 for "after"

20302 Chevrolet Trailblazer
Includes make-models: CG 18314 & 18316 (Chevrolet Trailblazer & EXT, GMC Envoy, Olds Bravada, Buick Rainier, Isuzu Ascender)
FMVSS No. 201 certification: 2002
Produced: 2002-2005
Side air bags: non-VIN-decodable option in 2003-2004
Frontal air bags: always
Pretensioners: never
Major redesigns: none
Decision: exclude because no pre-FMVSS No. 201 SUVs

20312 Chevrolet Tahoe
Includes make-models: 20312, 20313 (Chevrolet Tahoe 4-door); 23312, 23313 (GMC Yukon 4-door); 23318 (GMC Denali)
Excludes make-models: 20310, 20311, 23310, 23311 (Tahoe and Yukon 2-door), because discontinued before FMVSS No. 201 certification; Cadillac Escalade, because in separate group 19312
FMVSS No. 201 certification: 2000
Produced: always
Side air bags: torso bags standard in 2000-2002, non-VIN-decodable option in 2003-2004
Frontal air bags: driver in 1996, dual in 1997-2004
Pretensioners: never
Major redesigns: 2000
Decision: exclude because FMVSS No. 201-certification coincided with standard torso bags

20322 GM Suburban
Includes make-models: 20322, 20323, 20326, 20327, 23322, 23323, 20326, 23327, 23323
Excludes make-models: Chevrolet Avalanche, because no pre-FMVSS No. 201 vehicles
FMVSS No. 201 certification: 2000
Produced: always
Side air bags: torso bags standard in 2000-2002, non-VIN-decodable option in 2003-2004
Frontal air bags: driver in 1996, dual in 1997-2004
Pretensioners: never
Major redesigns: 2000
Decision: exclude because FMVSS No. 201-certification coincided with standard torso bags

20330 Chevrolet Tracker
Includes make-models: 20330, 20331, 20332, 20333 (Chevrolet Tracker); 53310, 53311, 53312, 53313 (Suzuki Sidekick); 53330, 53331, 53332, 53333 (Suzuki Vitara); 53336, 53337 (Suzuki Grand Vitara)
FMVSS No. 201 certification: 1999
Produced: always
Side air bags: never
Frontal air bags: always
Pretensioners: never
Major redesigns: 1999
Decision: use 1996-1998 for "before" and 1999-2001 for "after"

20406 GM Astro/Safari Van
Includes make-models: 20404, 20405, 20406, 20407 (Chevrolet Astro); 23404, 23405, 23406, 23407 (GMC Safari)
FMVSS No. 201 certification: 2003
Produced: always
Side air bags: never
Frontal air bags: always
Pretensioners: never
Major redesigns: none
Decision: use 2000-2002 for "before" and 2003-2005 for "after"

20410 GM full-size vans
Includes make-models: 20410-20436, 20610-20636, 23410-23436, 23610-23636 with GVWR under 10,000 pounds
Excludes make-models: 20638, 23638 (Chevrolet & GMC Cutaway)
FMVSS No. 201 certification: 2003
Produced: always
Side air bags: never
Frontal air bags: 1997-2005
Pretensioners: never
Major redesigns: none
Decision: use 2000-2002 for "before" and 2003-2005 for "after"

20452 Chevrolet Venture
Includes make-models: CG 18408, 18409 (Chevrolet Venture, Oldsmobile Silhouette, Pontiac Trans Sport & Montana)
FMVSS No. 201 certification: 2002
Produced: 1997-2004
Side air bags: none in 1997, standard torso bags in 1998-2000; standard combination bags for drivers & torso bags for RF in 2001-2002; in 2003-2004, this is usually an option
Frontal air bags: always
Pretensioners: 1998-2004
Major redesigns: none
Decision: use 2001 for "before" and 2002 for "after" (only years consistent on side air bags)

21022 Oldsmobile Aurora
Includes make-models: 21022
FMVSS No. 201 certification: 2001
Produced: 1996-1999 and 2001-2003
Side air bags: torso bags standard in 2001-2003
Frontal air bags: always
Pretensioners: never
Major redesigns: 2001
Decision: exclude because FMVSS No. 201-certification coincided with standard torso bags

22018 Pontiac Grand Am
Includes make-models: 18018 (Buick Skylark), 21021 (Oldsmobile Alero), 22018 (Pontiac Grand Am)
FMVSS No. 201 certification: 1999
Produced: always
Side air bags: never
Frontal air bags: always
Pretensioners: never
Major redesigns: 1999
FMVSS No. 214 certification: 1997
Decision: use 1997-1998 for "before" and 1999-2000 for "after"

22020 Pontiac Grand Prix
Includes make-models: 22020
FMVSS No. 201 certification: 2003
Produced: always
Side air bags: a small proportion of VIN-identified head curtains in 2004-2005
Frontal air bags: always
Pretensioners: 2004-2005
Major redesigns: 1997
Decision: use 2002 for "before" and 2003 for "after"; for unrestrained occupants, use 2000-2002 for "before" and 2003-2005 for "after," excluding cars with head curtains (pretensioners not an issue)

22352 Pontiac Aztek
Includes make-models: 22352, 22353
FMVSS No. 201 certification: 2001
Produced: 2001-2004
Side air bags: combination bags for drivers and torso bags for RF standard or optional
Frontal air bags: always
Pretensioners: GM letters say 2001 only (?), cars.com says always
Major redesigns: none
Decision: exclude because no pre-FMVSS No. 201 SUVs

24005 Saturn L Series
Includes make-models: 24005, 24006
FMVSS No. 201 certification: 2003
Produced: 2000-2004
Side air bags: head curtains optional in 2001 and standard in 2002-2004
Frontal air bags: always
Pretensioners: never
Major redesigns: none
Decision: exclude because all FMVSS No. 201-certified cars are equipped with head-protection air bags

24362 Saturn Vue
Includes make-models: 24362, 24363
FMVSS No. 201 certification: 2002
Produced: 2002-2004
Side air bags: optional non-VIN-decodable head curtains in each year
Frontal air bags: always
Pretensioners: 2004 only
Major redesigns: none
Decision: exclude because no pre-FMVSS No. 201 SUVs

30040 Volkswagen Jetta
Includes make-models: 30040
FMVSS No. 201 certification: upon redesign to CG 30010 in 1999
Produced: always
Side air bags: torso bags standard on CG 30010 in 1999-2000 and head curtains plus torso bags in 2001-2004
Frontal air bags: always
Pretensioners: always
Major redesigns: 1999
Decision: exclude because FMVSS No. 201-certification coincided with standard torso bags

30042 Volkswagen Golf
Includes make-models: 30042
FMVSS No. 201 certification: Golf in 2002 and GTI in 2003
Produced: always
Side air bags: torso bags standard on CG 30010 in 1999-2000 and head curtains plus torso bags in 2001-2004
Frontal air bags: always
Pretensioners: always
Major redesigns: 1999
Decision: exclude because all FMVSS No. 201-certified cars are equipped with head-protection air bags

30046 Volkswagen Passat
Includes make-models: 30046
FMVSS No. 201 certification: 2002 (except possibly on W8 model)
Produced: always
Side air bags: torso bags standard in 1998-2000 and head curtains plus torso bags in 2001-2004
Frontal air bags: always
Pretensioners: always
Major redesigns: 1998
Decision: exclude because all FMVSS No. 201-certified cars are equipped with head-protection air bags

30047 Volkswagen New Beetle
Includes make-models: 30047
FMVSS No. 201 certification: 2002
Produced: 1998-2004
Side air bags: torso bags standard in 1998-2003 and combination bags in 2004
Frontal air bags: always
Pretensioners: always
Major redesigns: none
Decision: use 2000-2001 for "before" and 2002-2003 for "after"

32000 Audi
Includes make-models: all cars & LTVs with MAK2 = 32
FMVSS No. 201 certification: 2002 or 2003
Side air bags: side air bags with head protection standard by 2001
Decision: exclude because all FMVSS No. 201-certified cars are equipped with head-protection air bags

34000 BMW
Includes make-models: all cars & LTVs with MAK2 = 33 (Mini-Cooper), 34 (BMW)
FMVSS No. 201 certification: with head-protection air bags, except for convertibles
Decision: exclude because almost all FMVSS No. 201-certified cars are equipped with head-protection air bags

35039 Nissan Maxima
Includes make-models: 35039
FMVSS No. 201 certification: 2000
Produced: always
Side air bags: a moderate proportion of VIN-identified torso bags in 1998-1999; a moderate proportion of VIN-identified combination bags in 2000-2003; head curtains plus torso bags standard in 2004
Frontal air bags: always
Pretensioners: 2000-2004 (according to cars.com; manufacturer's letter implausibly says 1999 also)
Major redesigns: 2000, 2004
Decision: generally exclude because pretensioners were apparently added in 2000; however, for unrestrained occupants, use 1997-1999 for "before" and 2000-2002 for "after," excluding cars with side air bags

35043 Nissan Sentra
Includes make-models: 35043
FMVSS No. 201 certification: 2000
Produced: always
Side air bags: a small proportion of VIN-identified combination bags in 2000-2004
Frontal air bags: always
Pretensioners: 2000-2004
Major redesigns: 2000 (but without changing wheelbase, staying in same CG)
Decision: generally exclude because pretensioners were added in 2000; however, for unrestrained occupants, use 1997-1999 for "before" and 2000-2002 for "after," excluding cars with combination bags

35047 Nissan Altima
Includes make-models: 35047
FMVSS No. 201 certification: 1999
Produced: always
Side air bags: a small proportion of VIN-identified combination bags in 2000-2001; a small proportion of VIN-identified head curtains plus torso bags in 2002-2004
Frontal air bags: always
Pretensioners: 2000-2004
Major redesigns: 2002
FMVSS No. 214 certification: 1997
Decision: use 1998 for "before" and 1999 for "after"; for unrestrained occupants, use 1997-1998 for "before" and 1999-2001 for "after" (pretensioners not an issue), excluding cars with combination bags

35200 Nissan Frontier
Includes make-models: 35200-35205
FMVSS No. 201 certification: 2002
Produced: 1998-2004
Side air bags: never
Frontal air bags: dual air bags with on-off switches in conventional and king cabs, dual air bags in crew cabs
Pretensioners: 2000-2004
Major redesigns: none
Decision: use 2000-2001 for "before" and 2002-2004 for "after"

35302 Nissan Pathfinder
Includes make-models: 35302-35303 (Nissan Pathfinder), 58302-58303 (Infiniti QX4)
FMVSS No. 201 certification: 2003
Produced: always
Side air bags: side air bags with head protection a non-VIN-decodable option in 1999-2004 on Pathfinder, standard on QX4 in 2000-2003
Frontal air bags: 1997-2004
Pretensioners: 2000-2004
Major redesigns: none
Decision: exclude because an unidentifiable portion of FMVSS No. 201-certified SUVs are equipped with head-protection air bags

35322 Nissan Xterra
Includes make-models: 35322, 35323
FMVSS No. 201 certification: 2000
Produced: 2000-2004
Side air bags: optional non-VIN-decodable head curtains 2003-2004
Frontal air bags: always
Pretensioners: always
Major redesigns: none
Decision: exclude because no pre-FMVSS No. 201 SUVs

37031 Honda Civic
Includes make-models: 37031
Excludes: Civic Hybrid (CG = 37037, MM2 = 37031)
FMVSS No. 201 certification: 2001
Produced: always
Side air bags: a moderate proportion of non-VIN-identified torso bags in 2001; a small proportion of VIN-identified torso bags in 2000-2003
Frontal air bags: always
Pretensioners: 2001-2004
Major redesigns: 2001 (but without changing wheelbase, staying in same CG)
Decision: generally exclude because pretensioners were added in 2001; however, for unrestrained occupants, use 1999-2000 for "before" and 2002-2004 for "after," excluding cars with torso bags, also excluding entire MY 2001 (torso bags not VIN-decodable)

37032 Honda Accord
Includes make-models: 37032
FMVSS No. 201 certification: 1999
Produced: always
Side air bags: large proportions of VIN-identified as well as non-VIN-identified torso bags in 2000-2002; large proportions of VIN-identified head curtains with torso bags as well as torso bags only in 2003-2004
Frontal air bags: always
Pretensioners: 2001-2004
Major redesigns: 2-door in 1998, 4-door in 2003
Decision: use 1997-1998 for "before" and 1999 for "after"

37035 Honda S2000
Includes make-models: 37035 (S2000), 37037 (Insight)
FMVSS No. 201 certification: 2000
Produced: 2000-2004
Side air bags: never
Frontal air bags: always
Pretensioners: always
Major redesigns: none
Decision: exclude because no pre-FMVSS No. 201 cars

37302 Honda CR-V
Includes make-models: 37302, 37303
FMVSS No. 201 certification: 2002
Produced: 1997-2004
Side air bags: the vast majority had VIN-identified torso bags in 2002-2004
Frontal air bags: always
Pretensioners: 1998-2004
Major redesigns: 2002
Decision: exclude because FMVSS No. 201-certification coincided with a vast majority of torso bags

37322 Honda Passport
Includes make-models: 37322, 37323
Excludes: Honda Pilot (CG = 37302, MM2 = 37322, 37323)
FMVSS No. 201 certification: 1999
Produced: 1996-2002
Side air bags: never
Frontal air bags: always
Pretensioners: never
Major redesigns: 1998
Decision: use 1996-1997 for "before" and 1999-2000 for "after"; skip 1998 because this first year of the redesigned Passport may have already anticipated FMVSS No. 201, even though the regulatory phase-in of FMVSS No. 201 didn't start until 1999

37402 Honda Odyssey
Includes make-models: 37402
FMVSS No. 201 certification: 1999
Produced: always
Side air bags: torso bags standard in 2002-2004
Frontal air bags: always
Pretensioners: 1999-2004
Major redesigns: 1999
Decision: generally exclude because pretensioners were added in 1999; however, for unrestrained occupants, use 1996-1998 for "before" and 1999-2000 for "after"

38300 Isuzu Rodeo Sport
Includes make-models: 38300, 38301 (Amigo & Rodeo Sport)
FMVSS No. 201 certification: 2003 (see comment below)
Produced: 1998-2003
Side air bags: never
Frontal air bags: always
Pretensioners: never
Major redesigns: none
Decision: exclude because only about 200 MY 2003 SUVs were registered in the United States, and there are no FARS cases of them; in fact, I wouldn't be surprised if these "2003" vehicles were actually manufactured well before 9/1/2003 and not actually certified to FMVSS No. 201

38322 Isuzu Rodeo
Includes make-models: 38322, 38323
FMVSS No. 201 certification: 1999
Produced: always
Side air bags: never
Frontal air bags: always
Pretensioners: never
Major redesigns: 1998
Decision: use 1996-1997 for "before" and 1999-2001 for "after"; skip 1998 because this first year of the redesigned Rodeo may have already anticipated FMVSS No. 201, even though the regulatory phase-in of FMVSS No. 201 didn't start until 1999

38326 Isuzu Axiom
Includes make-models: 38326, 38327
FMVSS No. 201 certification: 2002
Produced: 2002-2004
Side air bags: never
Frontal air bags: always
Pretensioners: never
Major redesigns: none
Decision: exclude because no pre-FMVSS No. 201 SUVs

39000 Jaguar
Includes make-models: all cars with MAK2 = 39 except S type (39034)
FMVSS No. 201 certification: 2000, 2002 or 2003
Side air bags: side air bags with head protection standard by 2001
Decision: exclude XK coupe (39031) because all FMVSS No. 201-certified cars are equipped with head-protection air bags; exclude XJ sedan (39032) because only 1 FARS case of a FMVSS No. 201-certified car without head-protection air bags; exclude X type (39036) because no pre-FMVSS No. 201 cars

41035 Mazda Protege
Includes make-models: 41035
FMVSS No. 201 certification: 1999 (except cars with sliding roofs; according to Ward's about 3% of
    Proteges have sliding roofs; although not VIN-identifiable, they are a small enough percentage that
    they may be ignored)
Produced: 1996-2003
Side air bags: a small proportion of VIN-identified combination bags in 2001-2003
Frontal air bags: always
Pretensioners: 2001-2003
Major redesigns: 1999
FMVSS No. 214 certification: 1995
Decision: use 1997-1998 for "before" and 1999-2000 for "after"; for unrestrained occupants, use 1996-1998
    for "before" and 1999-2001 for "after," excluding cars with combination bags (pretensioners not an
    issue)

41045 Mazda Miata
Includes make-models: 41045
FMVSS No. 201 certification: 2003
Produced: always except 1998
Side air bags: never
Frontal air bags: always
Pretensioners: 2001-2005
Major redesigns: none
Decision: use 2001-2002 for "before" and 2003-2005 for "after"

41402 Mazda MPV
Includes make-models: 41402, 41403
FMVSS No. 201 certification: 2000
Produced: always except 1999
Side air bags: a very small proportion (less than 1% according to Ward's) of non-VIN-identified
    combination bags in 2001; a large proportion of VIN-identified combination bags in 2001-2004
Frontal air bags: always
Pretensioners: 2002-2004
Major redesigns: 2000
Decision: use 1996-1998 for "before" and 2000-2001 for "after," excluding vans with combination bags

42042 Mercedes C
Includes make-models: 42042
FMVSS No. 201 certification: 2001
Produced: always
Side air bags: torso bags standard in 1998-2000 and head curtains plus torso bags in 2001-2004
Frontal air bags: always
Pretensioners: always
Major redesigns: 2001
Decision: exclude because all FMVSS No. 201-certified cars are equipped with head-protection air bags

42043 Mercedes S
Includes make-models: 42043 (S class), 42046 (CL coupe)
FMVSS No. 201 certification: 2000
Produced: always
Side air bags: torso bags standard in 1998-1999 and head curtains plus torso bags in 2000-2004
Frontal air bags: always
Pretensioners: always
Major redesigns: 2000
Decision: exclude because all FMVSS No. 201-certified cars are equipped with head-protection air bags

42044 Mercedes SL
Includes make-models: 42044
FMVSS No. 201 certification: 2003
Produced: always
Side air bags: torso bags standard in 1997-2002 and combination bags in 2003-2004
Frontal air bags: always
Pretensioners: always
Major redesigns: 2003
Decision: exclude because all FMVSS No. 201-certified cars are equipped with head-protection air bags

42045 Mercedes SLK
Includes make-models: 42045
FMVSS No. 201 certification: 2002
Produced: 1998-2004
Side air bags: torso bags standard in 1998-2004
Frontal air bags: always
Pretensioners: always
Major redesigns: none
Decision: use 1999-2001 for "before" and 2002-2004 for "after"

42047 Mercedes CLK
Includes make-models: 42047
FMVSS No. 201 certification: 2003
Produced: 1998-2004
Side air bags: torso bags standard in 1998-2002 (coupe), 1998-2003 (CV); head curtains plus torso bags standard on coupe in 2003-2004; combination bags standard on CV in 2004
Frontal air bags: always
Pretensioners: always
Major redesigns: 2003 (coupe), 2004 (convertible)
Decision: exclude because there are only 3 FARS cases of FMVSS No. 201-certified cars without head-protection air bags

42048 Mercedes E
Includes make-models: 42048
FMVSS No. 201 certification: 2000
Produced: always
Side air bags: torso bags standard in 1997-1998 and head curtains plus torso bags in 1999-2004
Frontal air bags: always
Pretensioners: always
Major redesigns: 2003
Decision: exclude because all FMVSS No. 201-certified cars are equipped with head-protection air bags

42303 Mercedes ML
Includes make-models: 42303, 42307
FMVSS No. 201 certification: 2000
Produced: 1998-2004
Side air bags: torso bags standard in 1998-2001 and head curtains plus torso bags in 2002-2004
Frontal air bags: always
Pretensioners: always
Major redesigns: none
Decision: use 1998-1999 for "before" and 2000-2001 for "after"

45000 Porsche
Includes make-models: 45031 (911), 45040 (Boxster)
FMVSS No. 201 certification: 2000
Produced: always
Side air bags: torso bags standard in 1999, 2002-2004 (also 1998 on Boxster); combination bags standard in 2000-2001
Frontal air bags: always
Pretensioners: 2002-2004
Major redesigns: 1999
Decision: exclude because FMVSS No. 201-certification coincided with head-protection air bags

47000 Saab
Includes make-models: 47035 (9-3), 47036 (9-5)
FMVSS No. 201 certification: 2002 for 9-3 convertible and 9-5; 1999 for 9-3 other than convertible
Produced: 1999-2004
Side air bags: combination bags or head curtains with torso bags standard in all years
Decision: exclude because all cars equipped with head-protection air bags

48034 Subaru Legacy
Includes make-models: 48034
FMVSS No. 201 certification: 2000 for cars without sun-roofs; 2001 for cars with sun-roofs, a non-VIN-identifiable option in 34% of 2000 cars according to Ward's
Produced: always
Side air bags: moderate proportions of VIN-identified torso bags in 2000-2004
Frontal air bags: always
Pretensioners: 2000-2004
Major redesigns: 2000
Decision: generally exclude because pretensioners were added in 2000; however, for unrestrained occupants, use 1999 for "before" and 2001-2002 for "after," excluding cars with torso bags and skipping 2000 because 201 certification is non-VIN-identifiable

48038 Subaru Impreza
Includes make-models: 48038
FMVSS No. 201 certification: 2002
Produced: always
Side air bags: the vast majority had VIN-identified torso bags in 2002-2004
Frontal air bags: always
Pretensioners: 2002-2004
Major redesigns: 2002
Decision: exclude because FMVSS No. 201-certification coincided with a vast majority of torso bags and also with pretensioners

48303 Subaru Forester
Includes make-models: 48303
FMVSS No. 201 certification: conflicting information for 2002; one letter says no, the other letter says yes for SUVs without sun-roofs, a non-VIN-identifiable option in 28% of 2002 SUVs according to Ward's
Produced: 1998-2004
Side air bags: moderate proportions of VIN-identified torso bags in 2001-2002; combination bags standard in 2003-2004
Frontal air bags: always
Pretensioners: 2003-2004
Major redesigns: none
Decision: exclude because all SUVs known for sure to be FMVSS No. 201-certified are equipped with head-protection air bags

49032 Toyota Corolla
Includes make-models: 20032 (Chevrolet Prizm), 49032 (Toyota Corolla)
FMVSS No. 201 certification: 2003
Produced: always
Side air bags: a large proportion of VIN-identified torso bags in 1998 and small proportions in 1999-2004; small proportions with VIN-identified torso or torso+curtain bags in 2005
Frontal air bags: always
Pretensioners: 1998-2005
Major redesigns: 2003
Decision: use 2000-2002 for "before" and 2003-2005 for "after," excluding cars with torso or torso+curtain bags

49033 Toyota Celica
Includes make-models: 49033
FMVSS No. 201 certification: 2000
Produced: always
Side air bags: a small proportion of VIN-identified torso bags in 2000-2004
Frontal air bags: always
Pretensioners: 2000-2004
Major redesigns: 2000
FMVSS No. 214 certification: 1996
Decision: generally exclude because pretensioners were added in 2000; however, for unrestrained occupants, use 1996-1999 for "before" and 2001 for "after," excluding cars with torso bags (the unusual choice of model years is necessary to balance the sample, because 2000 sales far exceeded 1996-1999 combined)

49040 Toyota Camry
Includes make-models: 49040
FMVSS No. 201 certification: 1999
Produced: always
Side air bags: a small proportion of VIN-identified torso bags in 1998-2001; a small proportion of VIN-identified head curtains plus torso bags in 2002-2004
Frontal air bags: always
Pretensioners: 1998-2004
Major redesigns: 1997, 2002
FMVSS No. 214 certification: 1994
Decision: Camry presents unique problems because at least some of the modifications needed for FMVSS No. 201 certification were already implemented in the 1997 redesign (even though "certification" did not officially exist until the 1999 phase-in). This is not merely speculation. Our tests of 1998 and 1999 Camry show similar overall performance with improvement only on a few targets. Our tests likewise show substantial additional improvement built into the 2002 redesign. There are really 4 generations of head impact performance: 1994-1996, 1997-1998, 1999-2001, and 2002-2004. Moreover, in picking from these, we must exclude cars with side air bags and, when including belted occupants, beware that pretensioners were added in 1998.

49041 Toyota MR2
Includes make-models: 49041
FMVSS No. 201 certification: 2000
Produced: 2000-2004
Side air bags: never
Frontal air bags: always
Pretensioners: always
Major redesigns: none
Decision: exclude because no pre-FMVSS No. 201 cars

49043 Toyota Avalon
Includes make-models: 49043
FMVSS No. 201 certification: 2003
Produced: always
Side air bags: torso bags standard in 1998-2004; torso + curtain standard in 2005
Frontal air bags: always
Pretensioners: 1999-2005
Major redesigns: 2005
Decision: use 2001-2002 for "before" and 2003-2004 for "after"; do not use 2005 because they all have curtains

49044 Toyota Camry Solara
Includes make-models: 49044
FMVSS No. 201 certification: 2000
Produced: 1999-2004
Side air bags: varying proportions of torso bags, possibly with head curtains, in 1999-2004
Frontal air bags: always
Pretensioners: always
Major redesigns: 2004
Decision: exclude because no pre-FMVSS No. 201 cars

49045 Toyota Echo
Includes make-models: 49045
FMVSS No. 201 certification: 2002
Produced: 2000-2004
Side air bags: a very small proportion of VIN-identified torso bags in 2001-2004
Frontal air bags: always
Pretensioners: always
Major redesigns: none
Decision: use 2000-2001 for "before" and 2002-2004 for "after," excluding cars with torso bags

49046 Toyota Prius
Includes make-models: 49046
FMVSS No. 201 certification: 2003
Produced: 2001-2005
Side air bags: a small proportion of VIN-identified torso bags in 2001-2003; a large proportion of VIN-identified head curtains plus torso bags in 2004-2005
Frontal air bags: always
Pretensioners: always
Major redesigns: 2004
Decision: use 2001-2002 for "before" and 2003-2005 for "after," excluding cars with torso or torso+curtain bags

49200 Toyota Tacoma with conventional cab or xtracab
Includes make-models: 49200-49203
FMVSS No. 201 certification: 2002
Produced: always
Side air bags: never
Frontal air bags: driver air bags in 1996-1997, dual air bags with on-off switches in 1998-2004
Pretensioners: 1999-2004
Major redesigns: none
Decision: use 2000-2001 for "before" and 2002-2004 for "after"

49204 Toyota Tacoma with double cab
Includes make-models: 49204-49205
FMVSS No. 201 certification: 2002
Produced: 2001-2005
Side air bags: never
Frontal air bags: always
Pretensioners: always
Major redesigns: none
Decision: use 2001 for "before" and 2002 for "after"

49210 Toyota Tundra with conventional or access cab
Includes make-models: 49210-49213
FMVSS No. 201 certification: 2003
Produced: 2000-2005
Side air bags: a small proportion of non-identifiable torso + curtain bags in 2005
Frontal air bags: dual air bags with on-off switches
Pretensioners: always
Major redesigns: none
Decision: use 2001-2002 for "before" and 2003-2004 for "after"

49302 Toyota 4Runner
Includes make-models: 49302, 49303
FMVSS No. 201 certification: 2003
Produced: always
Side air bags: head curtains plus torso bags are a non-VIN-identifiable option in 2003-2004; 20% of the
 SUVs, according to Ward's, were equipped with them in MY 2003-2004
Frontal air bags: always
Pretensioners: 1999-2004
Major redesigns: 2003
Decision: exclude because a substantial, non-VIN-identifiable proportion of the FMVSS No. 201-certified
 SUVs are equipped with head-protection air bags

49313 Toyota Landcruiser
Includes make-models: 49313 (Toyota Landcruiser), 59313 (Lexus LX470)
FMVSS No. 201 certification: 2003
Produced: always
Side air bags: head curtains plus torso bags are a non-VIN-identifiable option on 2003-2004 Landcruisers
 (30% so equipped according to Ward's) and standard on 2003-2004 LX470
Frontal air bags: always
Pretensioners: 1998-2004
Major redesigns: 1998
Decision: exclude because FARS has no cases of FMVSS No. 201-certified SUVs without head-protection
 air bags

49322 Toyota RAV4
Includes make-models: 49322, 49323
FMVSS No. 201 certification: 2001
Produced: always
Side air bags: head curtains plus torso bags are a non-VIN-identifiable option in 2004
Frontal air bags: always
Pretensioners: 1998-2004
Major redesigns: 2001
Decision: use 1998-2000 for "before" and 2001-2003 for "after"

49342 Toyota Highlander
Includes make-models: 49342, 49343
FMVSS No. 201 certification: 2001, based on Toyota's letter (*Buying a Safer Car* is in error)
Produced: 2001-2004
Side air bags: head curtains plus torso bags are a non-VIN-identifiable option
Frontal air bags: always
Pretensioners: always
Major redesigns: none
Decision: exclude because no pre-FMVSS No. 201 SUVs

49352 Toyota Sequoia
Includes make-models: 49352, 49353
FMVSS No. 201 certification: 2001
Produced: 2001-2004
Side air bags: head curtains plus torso bags are a non-VIN-identifiable option in 2001-2003 and standard in 2004
Frontal air bags: always
Pretensioners: always
Major redesigns: none
Decision: exclude because no pre-FMVSS No. 201 SUVs

49402 Toyota Sienna
Includes make-models: 49402, 49403
FMVSS No. 201 certification: 2001
Produced: 1998-2004
Side air bags: torso bags are a non-VIN-identifiable option in 2001-2003, but according to Ward's nobody bought them; head curtains plus torso bags are a non-VIN-identifiable option in 2004
Frontal air bags: always
Pretensioners: always
Major redesigns: 2004
Decision: use 1998-2000 for "before" and 2001-2003 for "after"

51000 Volvo
Includes make-models: all cars & LTVs with MAK2 = 51
FMVSS No. 201 certification: 1999 or 2001
Side air bags: side air bags with head protection standard at or before FMVSS No. 201 certification
Decision: exclude because all FMVSS No. 201-certified cars are equipped with head-protection air bags

52034 Mitsubishi Galant
Includes make-models: 42034
FMVSS No. 201 certification: 1999 for cars without sun-roofs; 2002 for cars with sun-roofs, a non-VIN-identifiable option in 20-30% of 1999-2003 cars according to Ward's
Produced: always
Side air bags: torso bags are a non-VIN-identifiable option in 1999-2004; 11-22% of the cars, according to Ward's, were equipped with them
Frontal air bags: always
Pretensioners: 2004 only
Major redesigns: 2004
Decision: exclude because substantial percentages in 1999-2001 have unknown FMVSS No. 201 status (sun-roofs) and/or unknown side air bags

52037 Mitsubishi Eclipse
Includes make-models: 52037
FMVSS No. 201 certification: 2000
Produced: always
Side air bags: in 2000-2002, there are no side air bags on about half the cars, and they are a non-VIN-identifiable option on the other half; a small proportion of VIN-identified torso bags in 2003-2004
Frontal air bags: always
Pretensioners: never
Major redesigns: 2000
Decision: use 1998-1999 for "before" and 2000-2002 for "after," limiting to cars known not to have torso bags

52040 Mitsubishi Diamante
Includes make-models: 52040
FMVSS No. 201 certification: 2003
Produced: 1997-2004
Side air bags: never
Frontal air bags: always
Pretensioners: 1999-2004
Major redesigns: none
Decision: use 2002 for "before" and 2003-2004 for "after"

52046 Mitsubishi Lancer
Includes make-models: 52046
FMVSS No. 201 certification: 2002
Produced: 2002-2004
Side air bags: torso bags are a non-VIN-identifiable option on some subseries
Frontal air bags: always
Pretensioners: always
Major redesigns: none
Decision: exclude because no pre-FMVSS No. 201 cars

52333 Mitsubishi Montero
Includes make-models: 52333
Excludes make-models: 52336, 52337 (Montero Sport)
FMVSS No. 201 certification: 2001
Produced: always
Side air bags: torso bags standard 2001-2004
Frontal air bags: always
Pretensioners: 2002-2004
Major redesigns: 2001
Decision: exclude because FMVSS No. 201-certification coincided with standard torso bags

52336 Mitsubishi Montero Sport
Includes make-models: 52336, 52337
Excludes make-models: 52333 (Montero)
FMVSS No. 201 certification: 2003
Produced: 1997-2004
Side air bags: never
Frontal air bags: always
Pretensioners: for drivers in 2001-2002, for drivers and RF in 2003-2004
Major redesigns: none
Decision: use 2002 for "before" and 2003-2004 for "after," but in analyses including belted occupants exclude RF passengers because pretensioners were added for them in 2003

53033 Suzuki Esteem & Aerio
Includes make-models: 53032 (Esteem), 53033 (Aerio)
FMVSS No. 201 certification: never on Esteem, always on Aerio
Produced: Esteem, 1996-2002; Aerio, 2002-2005 (i.e., both sold in 2002)
Side air bags: never
Frontal air bags: always
Pretensioners: never on Esteem, always on Aerio
Major redesigns: mid-2002 (Esteem discontinued and Aerio introduced on same wheelbase)
Decision: generally exclude because pretensioners were added on Aerio; however, for unrestrained occupants, use 2000-2002 Esteem for "before" and 2002-2004 Aerio for "after"

53342 Suzuki XL-7
Includes make-models: 53342, 53343
FMVSS No. 201 certification: 2001
Produced: 2001-2004
Side air bags: never
Frontal air bags: always
Pretensioners: always
Major redesigns: none
Decision: exclude because no pre-FMVSS No. 201 SUVs

54033 Acura NSX
Includes make-models: 54033
FMVSS No. 201 certification: 2002
Produced: always
Side air bags: never
Frontal air bags: always
Pretensioners: always
Major redesigns: none
Decision: use 1999-2001 for "before" and 2002-2004 for "after"

54035 Acura TL
Includes make-models: 54035
FMVSS No. 201 certification: 1999
Produced: always
Side air bags: torso bags standard in 2000-2003, head curtains plus torso bags standard in 2004
Frontal air bags: always
Pretensioners: 2002-2004
Major redesigns: 1999, 2004
FMVSS No. 214 certification: 1996
Decision: use 1996-1998 for "before" and 1999 for "after"

54036 Acura RL
Includes make-models: 54036
FMVSS No. 201 certification: 2003
Produced: always
Side air bags: torso bags standard in 1999-2004, curtain plus torso standard in 2005
Frontal air bags: always
Pretensioners: always
Major redesigns: 2005
Decision: use 2001-2002 for "before" and 2003-2004 for "after"

54037 Acura CL
Includes make-models: 54037
FMVSS No. 201 certification: 2001
Produced: 1997-1999, 2001-2003
Side air bags: torso bags standard 2001-2003
Frontal air bags: always
Pretensioners: 2001-2003
Major redesigns: none
Decision: exclude because FMVSS No. 201-certification coincided with standard torso bags and pretensioners

54038 Acura RSX
Includes make-models: 54038
FMVSS No. 201 certification: 2002
Produced: 2002-2004
Side air bags: torso bags standard
Frontal air bags: always
Pretensioners: always
Major redesigns: none
Decision: exclude because no pre-FMVSS No. 201 cars

54323 Acura MDX
Includes make-models: 54323
FMVSS No. 201 certification: 2001
Produced: 2001-2004
Side air bags: torso bags standard in 2001-2003, head curtains plus torso bags standard in 2004
Frontal air bags: always
Pretensioners: always
Major redesigns: none
Decision: exclude because no pre-FMVSS No. 201 SUVs

55033 Hyundai Sonata
Includes make-models: 55033
FMVSS No. 201 certification: 1999
Produced: always
Side air bags: torso bags standard 1999-2000, combination bags standard 2001-2004
Frontal air bags: always
Pretensioners: 1999-2004
Major redesigns: none
Decision: exclude because FMVSS No. 201-certification coincided with standard torso bags and pretensioners

55035 Hyundai Elantra
Includes make-models: 55035
FMVSS No. 201 certification: 2001
Produced: always
Side air bags: combination bags standard 2001-2004
Frontal air bags: always
Pretensioners: 1999-2004
Major redesigns: 2001
Decision: exclude because FMVSS No. 201-certification coincided with head-protection air bags

55036.3 Hyundai Accent 3HB
Includes make-models: 55036, BOD2 = 3 (body-type code for 3-door hatchback)
FMVSS No. 201 certification: 2000
Produced: always
Side air bags: combination bags are a non-VIN-identifiable option in 2003, standard in 2004
Frontal air bags: always
Pretensioners: 2000-2004
Major redesigns: 2000
Decision: generally exclude because pretensioners were added in 2000; however, for unrestrained occupants, use 1997-1999 for "before" and 2000-2002 for "after"

55036.4 Hyundai Accent 4SD
Includes make-models: 55036, BOD2 = 4 (body-type code for 4-door sedan)
FMVSS No. 201 certification: unknown: Hyundai letters say yes in 2000, no in 2001-2002
Produced: always
Side air bags: combination bags are a non-VIN-identifiable option in 2003, standard in 2004
Frontal air bags: always
Pretensioners: 2000-2004
Major redesigns: 2000
Decision: exclude because of uncertainty when they were FMVSS No. 201 certified

55037 Hyundai Tiburon
Includes make-models: 55037
FMVSS No. 201 certification: 2002 (in effect 2003, because Tiburon produced during 2002 were sold as 2003 models)
Produced: 1997-2001, 2003-2004
Side air bags: combination bags standard 2003-2004
Frontal air bags: always
Pretensioners: 2000-2004
Major redesigns: 2003
Decision: exclude because FMVSS No. 201-certification coincided with head-protection air bags

55038 Hyundai XG350
Includes make-models: 55038
FMVSS No. 201 certification: 2001 (except not until 2003 with sun-roof)
Produced: 2001-2004
Side air bags: combination bags standard
Frontal air bags: always
Pretensioners: always
Major redesigns: none
Decision: exclude because no pre-FMVSS No. 201 cars

55302 Hyundai Santa Fe
Includes make-models: 55302, 55303
FMVSS No. 201 certification: 2001
Produced: 2001-2004
Side air bags: combination bags a non-VIN-identifiable option in 2002, standard in 2003-2004
Frontal air bags: always
Pretensioners: always
Major redesigns: none
Decision: exclude because no pre-FMVSS No. 201 SUVs

58032 Infiniti Q45
Includes make-models: 58032
FMVSS No. 201 certification: 2002
Produced: always
Side air bags: torso bags a VIN-identifiable option in 1997, standard in 1998; combination bags standard in 1999-2001; head curtains plus torso bags standard in 2002-2004
Frontal air bags: always
Pretensioners: always
Major redesigns: 1997, 2002
Decision: exclude because all FMVSS No. 201-certified cars are equipped with head-protection air bags

58035 Infiniti I30
Includes make-models: 58035 (I30), 58036 (I35)
FMVSS No. 201 certification: 2000
Produced: always
Side air bags: torso bags standard in 1998-1999; combination bags standard in 2000-2004
Frontal air bags: always
Pretensioners: 2000-2004 (according to cars.com; manufacturer's letter implausibly says 1999 also)
Major redesigns: 2000
Decision: exclude because FMVSS No. 201-certification coincided with head-protection air bags

59031 Lexus ES
Includes make-models: 59031
FMVSS No. 201 certification: 1999
Produced: always
Side air bags: torso bags standard in 1998-2001; head curtains plus torso bags standard in 2002-2004
Frontal air bags: always
Pretensioners: 1997-2004
Major redesigns: 1997, 2002
FMVSS No. 214 certification: 1994
Decision: Lexus ES presents unique problems because at least some of the modifications needed for FMVSS No. 201 certification were already implemented in the 1997 redesign (even though "certification" did not officially exist until the 1999 phase-in). This is not merely speculation. Our tests of 1998 and 1999 Camry (a "twin") show similar overall performance with improvement only on a few targets. There are really 3 generations of head impact performance: 1994-1996, 1997-1998 and 1999-2001. The 2002-2004 cars need not be considered because head curtains were standard. Use 1995-1996 for "before" and 1997 for "after" the 1997 transition (unrestrained only, because pretensioners added in 1997). Use 1998 for "before" and 1999 for "after" the 1999 transition.

59032 Lexus LS
Includes make-models: 59032
FMVSS No. 201 certification: 2001
Produced: always
Side air bags: torso bags standard in 1997-2000; head curtains plus torso bags standard in 2001-2004
Frontal air bags: always
Pretensioners: always
Major redesigns: 2001
Decision: exclude because FMVSS No. 201-certification coincided with head-protection air bags

59033 Lexus SC
Includes make-models: 59033
FMVSS No. 201 certification: 2001 (in effect 2002, because SC produced during 2001 were sold as 2002 models)
Produced: 1996-2000, 2002-2004
Side air bags: torso bags standard in 2002-2004
Frontal air bags: always
Pretensioners: always
Major redesigns: 2002
Decision: exclude because FMVSS No. 201-certification coincided with torso bags

59034 Lexus GS
Includes make-models: 59034
FMVSS No. 201 certification: 1999
Produced: always
Side air bags: torso bags standard in 1998-2000; head curtains plus torso bags standard in 2001-2004
Frontal air bags: always
Pretensioners: always
Major redesigns: 1998
Decision: exclude; should not use 1998 because this first year of the redesigned Lexus GS may have already anticipated FMVSS No. 201, even though the regulatory phase-in of FMVSS No. 201 didn't start until 1999; should not use pre-1998 because of the transition to torso bags in 1998

59035 Lexus IS
Includes make-models: 59035
FMVSS No. 201 certification: 2003
Produced: 2001-2004
Side air bags: torso bags standard in 2001; head curtains plus torso bags standard in 2002-2004
Frontal air bags: always
Pretensioners: always
Major redesigns: none
Decision: exclude because all FMVSS No. 201-certified cars are equipped with head-protection air bags

59332 Lexus RX300
Includes make-models: 59332, 59333 (RX 300); 59342, 59343 (RX 330)
FMVSS No. 201 certification: 2003
Produced: 1999-2004
Side air bags: torso bags standard in 1999-2003; head curtains plus torso bags standard in 2004
Frontal air bags: always
Pretensioners: always
Major redesigns: 2004
Decision: use 2002 for "before" and 2003 for "after"

62307 Land Rover Discovery
Includes make-models: 62303 (Discovery), 62307 (Discovery II)
FMVSS No. 201 certification: never on Discovery, always on Discovery II
Produced: Discovery, 1996-1999; Discovery II, 1999-2004 (i.e., both sold in 1999)
Side air bags: never
Frontal air bags: always
Pretensioners: never on Discovery, always on Discovery II
Major redesigns: mid-1999 (Discovery discontinued and Discovery II introduced on same wheelbase)
Decision: generally exclude because pretensioners were added on Discovery II; however, for unrestrained occupants, use 1996-1999 Discovery for "before" and 1999-2001 Discovery II for "after"

62313 Land Rover Range Rover
Includes make-models: 62313
FMVSS No. 201 certification: 2003
Produced: always
Side air bags: torso bags optional or standard in 1999-2002; head curtains plus torso bags standard in 2003-2004
Frontal air bags: always
Pretensioners: 2000-2004
Major redesigns: 2003
Decision: exclude because FMVSS No. 201-certification coincided with head-protection air bags

62341 Land Rover Freelander
Includes make-models: 62341, 62343
FMVSS No. 201 certification: 2002
Produced: 2002-2004
Side air bags: never
Frontal air bags: always
Pretensioners: always
Major redesigns: none
Decision: exclude because no pre-FMVSS No. 201 SUVs

63031 Kia Sephia
Includes make-models: 63031
FMVSS No. 201 certification: 1999
Produced: 1996-2001
Side air bags: never
Frontal air bags: always
Pretensioners: 2001 only
Major redesigns: 1998
FMVSS No. 214 certification: 1996
Decision: use 1996-1997 for "before" and 1999 for "after"; skip 1998 because this first year of the redesigned Sephia may have already anticipated FMVSS No. 201, even though the regulatory phase-in of FMVSS No. 201 didn't start until 1999

63032 Kia Rio
Includes make-models: 63032
FMVSS No. 201 certification: 2001
Produced: 2001-2004
Side air bags: never
Frontal air bags: always
Pretensioners: always
Major redesigns: none
Decision: exclude because no pre-FMVSS No. 201 cars

63033 Kia Spectra
Includes make-models: 63033
FMVSS No. 201 certification: almost certainly 2000; Kia's 2001 letter says yes, whereas their 2000 letter doesn't even mention Spectra, but it is hard to believe that the 2000 Spectra didn't already comply, especially since Sephia, built on the same wheelbase, already complied in 1999
Produced: 2000-2004
Side air bags: head curtains plus torso bags standard on mid-2004 redesigned Spectra (CG 63009)
Frontal air bags: always
Pretensioners: 2001-2004
Major redesigns: mid-2004
Decision: exclude because no pre-FMVSS No. 201 cars

63034 Kia Optima
Includes make-models: 63034
FMVSS No. 201 certification: 2001
Produced: 2001-2004
Side air bags: combination bags standard
Frontal air bags: always
Pretensioners: always
Major redesigns: none
Decision: exclude because no pre-FMVSS No. 201 cars

63300 Kia Sportage
Includes make-models: 63300, 63301, 63302, 63303
FMVSS No. 201 certification: 2002
Produced: 1996-2002
Side air bags: never
Frontal air bags: none in 1996, driver in 1997, dual in 1998-2002
Pretensioners: never
Major redesigns: none
Decision: use 2001 for "before" and 2002 for "after"

63402 Kia Sedona
Includes make-models: 63402
FMVSS No. 201 certification: 2002
Produced: 2002-2004
Side air bags: never
Frontal air bags: always
Pretensioners: always
Major redesigns: none
Decision: exclude because no pre-FMVSS No. 201 vans

64031 Daewoo Lanos
Includes make-models: 64031
FMVSS No. 201 certification: 2002
Produced: 1999-2002
Side air bags: never
Frontal air bags: always
Pretensioners: 2000-2002
Major redesigns: none
Decision: use 2001 for "before" and 2002 for "after"

64032 Daewoo Nubira
Includes make-models: 64032
FMVSS No. 201 certification: 2001
Produced: 1999-2002
Side air bags: never
Frontal air bags: always
Pretensioners: always
Major redesigns: none
Decision: use 1999-2000 for "before" and 2001-2002 for "after"

64033 Daewoo Leganza
Includes make-models: 64033
FMVSS No. 201 certification: 1999
Produced: 1999-2002
Side air bags: never
Frontal air bags: always
Pretensioners: always
Major redesigns: none
Decision: exclude because no pre-FMVSS No. 201 cars

DOT HS 811 538
November 2011

U.S. Department
of Transportation
**National Highway
Traffic Safety
Administration**